Automobile Transmission Systems

E. B. Weston

Newnes Technical Books

Newnes Technical Books
is an imprint of the Butterworth Group
which has principal offices in
London, Sydney, Toronto, Wellington, Durban and Boston

First published 1976
Reprinted 1979, 1981

ISBN 0 408 00184 4

Printed in England by Hazell, Watson & Viney Ltd., Aylesbury, Bucks

CONTENTS

PREFACE

This book, which is one of the Newnes-Butterworths Questions and Answers series, deals with the principles, operation and servicing of motor car transmissions. It is intended to be of assistance to owner-drivers, garage mechanics and others interested in the maintenance and repair of automobile transmissions in general.

The book is divided into eleven chapters, covering transmission systems from the engine to the driving force between tyre and road. The introduction defines a number of systems in use. This is followed by a study of the way in which gearing and gear ratios are determined. Friction type clutches are then examined in a variety of forms, including those with electrical or magnetic means of application. Alternative coupling systems in the shape of fluid flywheels and torque converters are then described and discussed. This chapter is followed by positive engagement and overrunning clutches. The subject of gearboxes opens with the layshaft type box, passing through sliding mesh and constant mesh to the synchromesh arrangements. Gear selector mechanisms conclude the topic of gearboxes and the book moves on to the subject of overdrives, as a separate unit and integral with the gearbox. Torque splitting devices for use on four-wheel drive cars are included at this stage.

Automatic transmissions are discussed in terms of basic gearsets and the means of selection, without details of individual application as to the hydraulic circuitry. Drive arrangements together with types of universal joints and propeller shaft assemblies bring the book to the subject of rear axles. Here the basic construction of 'live' axles in various forms is discussed, along with two-speed and double-reduction axles. Limited-slip and transaxles are also defined. The importance of the selection of tyres to be fitted to cars, along with the legal requirements and other considerations such as aquaplaning, bring the book to a close.

E.B.W.

Note on

SI (SYSTÈME INTERNATIONALE) UNITS

This System of International Units has been adopted by the UK, along with many other countries. The basic units have evolved by the work of a number of conferences over a long period of time. The foot, pound, second, system, measuring length, mass (force, as the poundal), and time, respectively, was in use before 1862, when the British Association appointed a committee to study units. This resulted in the C.G.S. system of 1872. The M.K.S. system of 1900, and the M.K.S.A. or Giorgi system of 1950, are now replaced by the SI system decided upon in 1960.

In this system, length is based on the metre (m), mass on the kilogram (kg), force on the newton (N), time on the second (s), electrical current on the ampere (A), thermodynamic temperature on the kelvin (K), and luminous intensity on the candela (cd). In all the systems mentioned above, each unit 'interlocks' to form a 'coherent' system; that is, multiplication of one unit with another (in the same system) will produce a derived unit that is still unity. For example:

$$1 \text{ newton} \times 1 \text{ metre} = 1 \text{ newton metre}$$

Not all of the above basic units are used in this book, but SI units, basic and derived, are used throughout the text, without reference to Imperial units. As conversion factors may help some readers to assess or compare values, the following list is appended.

Length.	Metre, m	1 mile	= 1609.34 m
			= 1.6 km
		1 yard	= 0.9144 m
		1 foot	= 0.3048 m
		1 inch	= 0.0254 m
Mass.	Kilogram, kg	1 ton	= 1016.05 kg
		1 pound	= 0.453 592 kg
Time.	Second, s	basic unit of time	

Force. Newton, N

The newton is that unit of force, which, applied to a mass of 1 kilogram, gives it an acceleration of 1 metre per second per second

1 N = 0.225 lbf
1 kgf = 9.806 N

Energy, work, heat. Joule, J

This is the work done, when the point of application of a force of 1 newton is displaced through the distance of one metre in the direction of the force. As a unit of heat, it is the direct thermal equivalent of the defined amount of mechanical energy and is thus independent of conversion factors

1 J = 0.737 56 ft lbf

Power. Watt, W

The watt is the unit of power equal to one joule per second, 1 J/s

1 kW = 1.341 h.p.
= 737.56 ft/lbf/s
1 h.p. = 745.7 W
1 metric h.p. = 735.5 W

Rotational frequency. Revolutions per second, rev/s, or revolutions per minute, rev/min, may be used

$$\frac{\text{rev/min}}{60} = \text{rev/s}$$

Some derived units used in the text

Velocity. In metres per second (m/s) or kilometres per hour (km/h).

Acceleration. In metres per second per second, usually written m/s^2. Note that this is a 'shorthand' way of writing, and does not indicate 'time squared'.

Torque. In newton metres; a force in newtons × a distance in metres (N m).

1

INTRODUCTION

What parts of the car make up the transmission system?

All the units that connect the engine with the road wheels, namely, the clutch, gearbox, propeller shaft and rear axle, as in Fig. 1. This arrangement was first introduced by Louis Renault in 1898. The system became very popular with other car manufacturers and through the years up to the 1940s it could be referred to as the 'orthodox' system. This layout is classified as 'front-engine rear-wheel drive'.

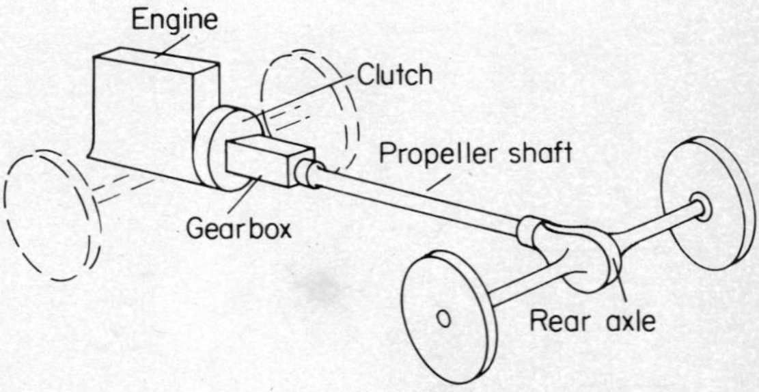

Fig. 1. Front-engine rear-wheel drive

What other two-wheel drive arrangements are possible?

Front-engine front-wheel drive, and rear-engine rear-wheel drive. Fig. 2 shows an arrangement using an in-line engine as used in the Triumph 1300. The position of the final drive and indeed the angle of the engine can vary from manufacturer to manufacturer. Fig. 3 shows the same basic arrangement but

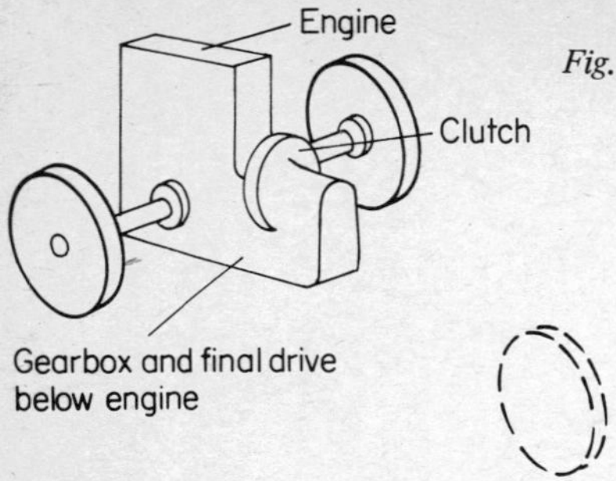

Fig. 2. Front (in-line) engine, front-wheel drive

with a transverse engine as used in the Mini and other B.L.M.C. cars. Fig. 4 illustrates the rear-engine rear-wheel drive arrangement and again the position of the engine and final drive varies from manufacturer to manufacturer. The Volkswagen, as an example, uses an air-cooled flat-four engine, while Renault, Fiat and others favour the vertical in-line engine.

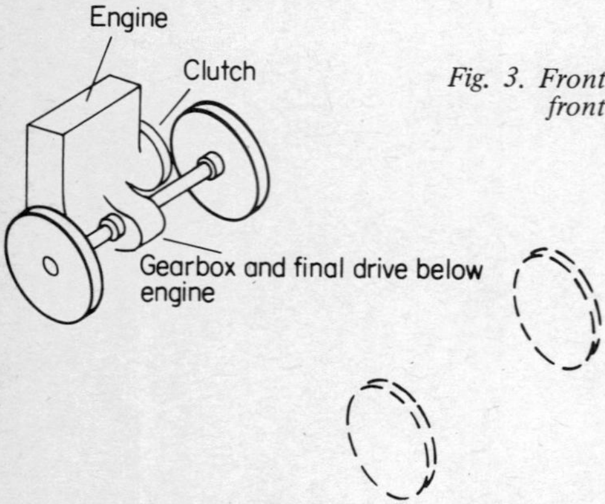

Fig. 3. Front (transverse) engine, front-wheel drive

Is a four-wheel drive possible?

Yes, a four-wheel drive system can be found on the Jensen, Range Rover, and special purpose-built cars for the Police, Ambulance, and other essential services. Fig. 5 gives an impression of a four-wheel drive, using a torque splitting device behind the gearbox, to provide a driving effort at front and rear wheels in proportion to the load carried on each axle.

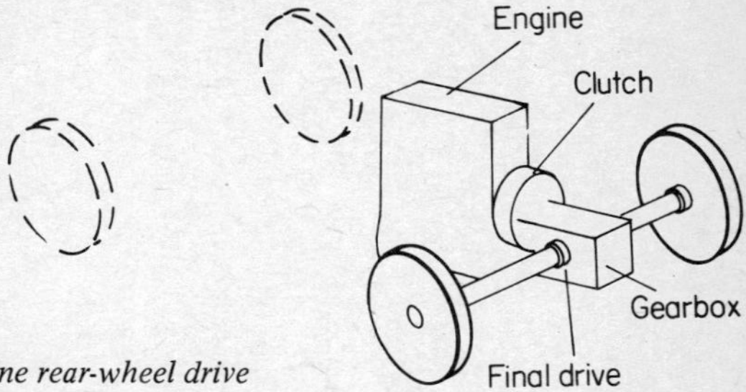

Fig. 4. Rear-engine rear-wheel drive

Why the departure from the 'orthodox' system in Fig. 1?

The main reason for the designs in Figs. 2, 3, and 4 is the elimination of the propeller shaft and its tunnel, through the car body, enabling a compact four-seater to be constructed with a considerable reduction in overall dimensions.

Four-wheel drives contribute to safety at speed, the torque splitting device proportioning the driving force in terms of axle loading for maximum adhesion with the road.

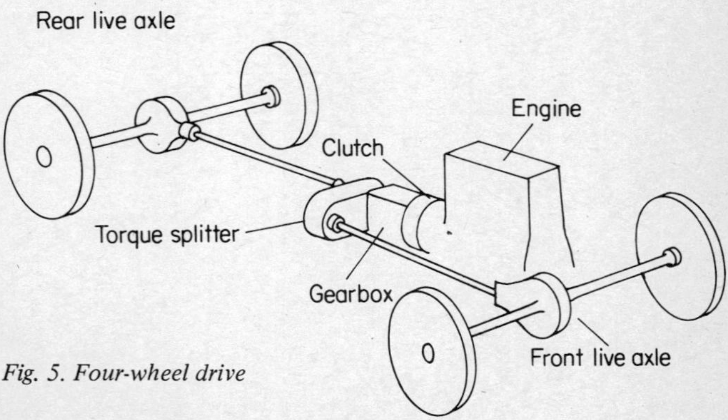

Fig. 5. Four-wheel drive

2

TRANSMISSION OF MOTION AND POWER

What determines the performance of a vehicle?

The performance of a vehicle depends on the power available at the engine crankshaft. The power at the road wheels will depend on the transmission losses and the resistances to motion due to conditions prevailing at the time. The opposition to motion of the vehicle will be some combination of the rolling resistance, air resistance, gradient resistance, and the force that may be required to accelerate the vehicle.

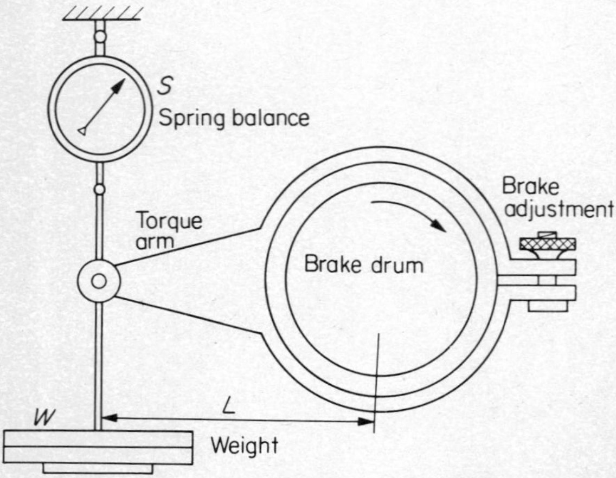

Fig. 6. Prony brake

How is the power of an engine measured?

Power is measured by coupling the engine to an absorption dynamometer, which is an apparatus capable of absorbing the power developed by the engine.

A simple dynamometer, such as the Prony brake (see Fig. 6) will illustrate the principle involved. The engine is coupled direct to the brake drum so that the power given by the engine, at a given speed and throttle opening, can be balanced by the torque arm L, and the force $W-S$.

L = Length of torque arm in metres
W = Force on torque arm in newtons
S = Force on spring balance in newtons
N = Revolutions per minute
Then T (brake torque) $= (W-S) \times L$ newton metres
Brake power = Torque $\times$ Angle turned through in rad/s

$$= T \times \frac{2\pi N}{60} \text{ joules per second (J/s)}$$

$$\text{or } \frac{2\pi NT}{60 \times 1000} \text{ kW}$$

2π radians = 1 revolution

Fig. 7. Characteristic speed/power curve for petrol engine

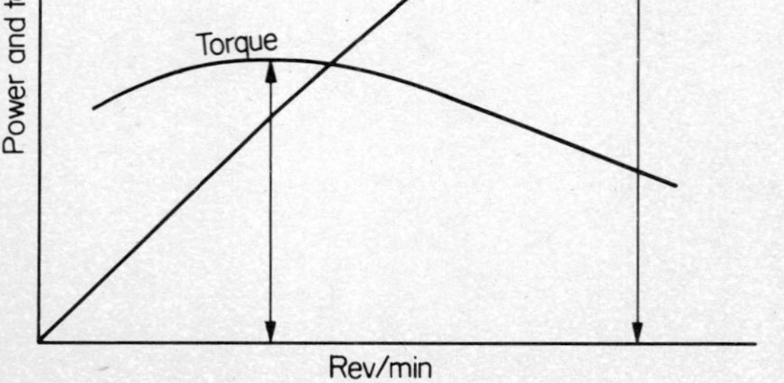

How is the power of an engine recorded?

If the engine speed is varied by the brake load while the throttle is set wide open, and the two variables recorded, a speed/power curve can be drawn (Fig. 7). Other tests can be carried out at three-quarter or half-throttle if required by the manufacturer. Other forms of dynamometer may be used to carry out similar tests.

What is rolling resistance?

The rolling resistance of any vehicle is the force necessary to maintain a slow steady speed, say 1 km/h, on a level road. Its value depends on the tyres fitted to the car, the nature of the road surface and the arrangements of the vehicle drive. The resistance is measured in newtons and may be expressed as a percentage of the total downwards force created by the vehicle itself.

Let R_r = Rolling resistance.

What is gradient resistance?

This is the additional resistance arising when the vehicle is climbing an incline (or hill) whose gradient is the ratio between the elevation a, and the distance travelled along the road b (Fig. 8). The gradient is represented by $\frac{a}{b} = \sin \theta$.

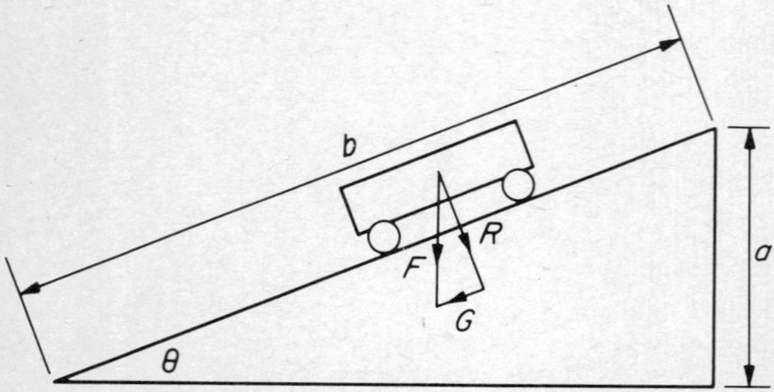

Fig. 8. Gradient resistance

The force diagram shows the downwards force F on the vehicle by the action of gravity. R is the reaction of the vehicle at right-angles to the road surface, and G is the force acting down the gradient.

Gradient resistance (R_g) = Direct downwards force × sin θ

$= F \times \sin\theta$

Forces opposing motion up the incline = Gradient resistance + Rolling resistance

$= R_g + R_r$

If a vehicle, standing on a gradient, is free to move when the handbrake is released, movement down the incline will take place if gradient resistance minus rolling resistance is positive in value.

Force acting down the gradient $= F \sin\theta - R_r$.

What is air resistance?

Air resistance is calculated from the formula $R_a = kV^2A$. The resistance increases with the square of the vehicle's speed. Maximum resistance will occur when the vehicle is travelling head-on into the wind. Side forces will vary in intensity as the vehicle changes direction relative to the air stream.

In the formula $R_a = kV^2A$

R_a	=	Air resistance in newtons
k	=	Constant depending on shape of vehicle
V	=	Velocity of vehicle in km/h
A	=	Frontal area of vehicle in m^2.

For definite values of k, wind tunnel tests are essential, but for approximate calculations values of k may be taken as:

0.044	for a lorry or bus
0.032	for a modern motor car
0.022	for a sports car
0.011	for a streamlined racing vehicle.

What force is available for acceleration?

All the above resistances, individual or in any combination one with the others, oppose the motion of the vehicle. The driving force must be of sufficient value to neutralise the opposing forces, and if acceleration is required an additional force over and above any resistance combination will have to be provided. Additional force required to produce acceleration (in newtons)

$$= \text{Mass (kg)} \times \text{Acceleration (m/s}^2\text{)}$$

$$F_a = \text{Mass} \times \text{Acceleration.}$$

The maximum force required to drive the car could be

$$R_r + R_9 + R_a + \tfrac{1}{2}F_a = \text{Total tractive resistance.}$$

Half the accelerating force is added as this is the average force acting during any acceleration period.

What determines the power available at the road wheels?

The power available at the road wheels is equal to the power at the engine crankshaft (brake power) minus friction losses incurred in the transmission. The friction losses will vary between driving in the intermediate gears and direct drive through the gearbox in top.

Power available at the road wheels (in watts) = Brake power of engine (in watts) × Transmission efficiency as a percentage

$$\text{Power available} = \text{b.p.} \times E$$

where E = Transmission efficiency expressed as a percentage.

If the total tractive effort in newtons and the vehicle speed in m/s are known:

Power available at the road wheels (in watts) = Tractive effort (in newtons) × Vehicle speed (in m/s)

The power developed at the crankshaft of the engine for any given road speed can be found by working back from the tractive effort.

b.p. of engine (in watts) = Tractive effort (in newtons) × Vehicle speed (in m/s) ÷ Transmission efficiency

$$\text{b.p.} = \frac{T_e \times V}{E} \quad \text{watts}$$

What power is available for acceleration?

The tractive effort can be calculated and the power required at any given road speed can be determined. This enables a graph of power available and power required to be constructed on a speed basis for any set of conditions, or resistance combinations. For top gear on a level road the two curves are arranged as shown in Fig. 9. It should be noted that

Fig. 9. Power available for acceleration

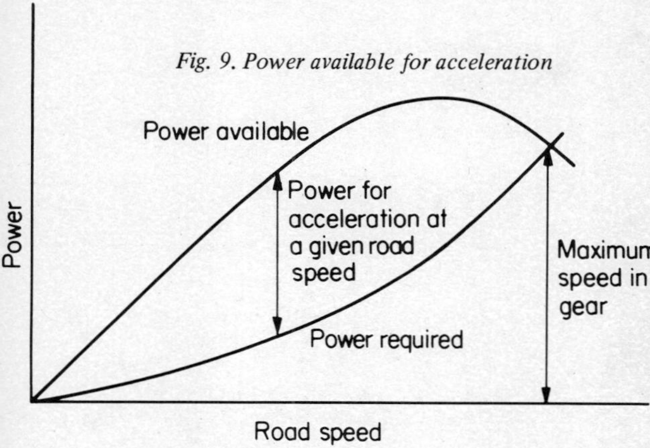

maximum power of the engine occurs at about 85% of the maximum road speed in km/h. This provides ample power for acceleration, as Power available − Power required = Power available for acceleration. If the curve of power required passed through the point of maximum power, then maximum road speed would occur at this position, and power available for acceleration would be reduced considerably.

What is the relationship between road speed and engine revolutions per minute?

The engine is tested for brake power by the makers and a speed/power graph is constructed to show variations in torque and brake power. From this graph (Fig. 7) the engine speed at maximum torque and maximum power can be obtained. The ratio of the speed at maximum torque to the speed at maximum power is called the speed ratio of the engine. This ratio is used to enable suitable gear ratios to be employed to drive the vehicle.

The road speed of the car will be determined by the revolutions per minute of the road wheels. Assuming that no slip takes place between the tyre and the road, the following relations will apply:

Let D = Effective diameter of wheel in mm
V = Speed of car in km/h
N = Revolutions of wheel per minute (rev/min).

$$\text{Then distance travelled per revolution} = \frac{\pi D}{1000}$$

$$\text{Distance travelled per minute} = \frac{\pi DN}{1000}$$

$$\text{and } V = \frac{\pi DN \times 60}{1000 \times 1000} \text{ km/h}$$

The ratio between engine rev/min and rear axle rev/min is called the overall gear ratio.

$$\text{Overall gear ratio} = \frac{\text{Engine rev/min}}{\text{Rear axle rev/min}}$$

What is the relationship between top gear performance on the level road and the rear axle ratio?

If the expected performance range of the vehicle is known, or is estimated, it can be equated to the critical speeds of the engine to produce a rear axle ratio from the calculations.

The lowest road speed in top gear is converted to rear axle rev/min and the ratio of speed at maximum torque (engine) to rear axle rev/min is calculated. The highest road speed in top gear is now converted to rear axle rev/min and the ratio of speed at maximum power (engine) to rear axle rev/min is calculated.

The average of the two results will provide a rear axle reduction ratio, suitable for further calculation.

Example: A Hillman 1600 has a rear axle ratio of 3.889:1. How does this compare with the calculated value to give a top gear performance of 60 to 120 km/h?

The tyre size of 5.50 × 13 provides an effective diameter of (2 × 5.50) + 13 = 24 in. Converting this to mm, 24 × 25.4 = 610 mm.

To find rev/min of axle shaft from road speed in km/h

transpose $V = \dfrac{\pi D n \times 60}{1000 \times 1000}$ to n.

$$n = \frac{V \times 1000 \times 1000}{\pi D \times 60}$$

To find rear axle rev/min at 120 km/h

$$n = \frac{120 \times 1000 \times 1000 \times 7}{22 \times 610 \times 60}$$

$$= 1044 \text{ rev/min}$$

$$\text{Rear axle ratio} = \frac{\text{Engine rev/min at max. power}}{\text{Rear axle rev/min at 120 km/h}}$$

$$= \frac{4100}{1044} = 3.927{:}1$$

To find rear axle rev/min at 60 km/h

$$n = \frac{60 \times 1000 \times 1000 \times 7}{22 \times 610 \times 60}$$

$$= 522 \text{ rev/min}$$

$$\text{Rear axle ratio} = \frac{2100}{522} = 4.023$$

Average = 3.975

How is the lowest gearbox ratio determined if the rear axle ratio is assumed to be 4:1?

This calculation will require further information, such as vehicle weight, coefficient of adhesion between tyre and road, the maximum torque available and the percentage of the vehicle weight carried by the rear axle.

Maximum torque = 117.7 N m
Unladen weight Hillman 1600 = 993 kg
Coefficient of adhesion = 0.8
Assume 60% of vehicle weight on rear axle
Weight on rear axle = 0.6 × 993 kg = 750 kg
Maximum tractive effort possible, without slip =

$$750 \times 0.8 \times 9.807 = 5866 \text{ N}$$

$$\text{Maximum torque at road wheel} = \frac{5866 \times 610}{2 \times 1000}$$

$$= 1788 \text{ N m}$$

$$\text{Overall gear ratio} = \frac{\text{Possible torque at road wheel}}{\text{Maximum engine torque}}$$

$$= \frac{1788}{117.7} = 15.20$$

Overall gear ratio = Gearbox ratio × Rear axle ratio

$$\text{Lowest gearbox ratio} = \frac{\text{Overall gear ratio}}{\text{Rear axle ratio}}$$

$$= \frac{15.20}{4}$$

$$= 3.55:1$$

What method is used to determine intermediate gear ratios?

The intermediate gear ratios are arranged in geometrical progression, with a common ratio between each stage.

Geometrical progress, as the graph (Fig. 10) shows, reduces the speeds sharply at the high end, and slowly at the bottom end, giving a range of gearing more suitable than a system based on arithmetic progression.

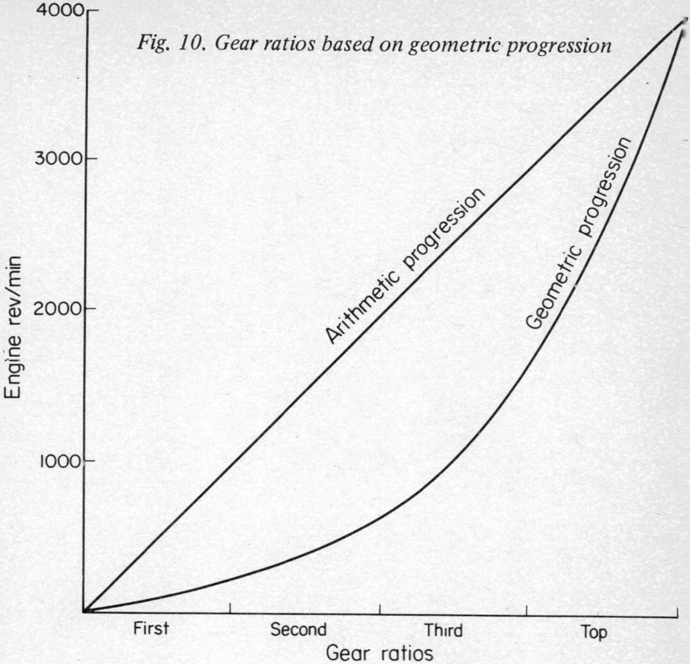

Fig. 10. Gear ratios based on geometric progression

Assuming lowest gearbox ratio to be 3.55:1 (as on page 20), the common ratio between the gears can be found if the gearbox is to have a direct drive in top. For a four-speed gearbox with a top gear ratio of 1:1

$$\text{Common ratio} = \sqrt[3]{\frac{1}{3.550}} = 0.6550$$

This common ratio should be equal to or greater than the speed ratio of the engine.

$$\text{Speed ratio} = \frac{2100}{4100} = 0.5121$$

To complete the gearbox ratio calculations, first gear and top gear are known, so that:

First gear	=	3.55:1
Second gear	=	$3.550 \times 0.6555 = 2.327{:}1$
Third gear	=	$3.845 \times 0.6555^2 = 1.1525{:}1$
Top gear	=	$3.550 \times 0.6555^3 = 1{:}1$

The calculated gear ratios are sometimes referred to as 'ideal' gear ratios, because they invariably have to be modified to accommodate all the necessary wheels on two parallel shafts, as in the layshaft type gearbox.

For the Hillman 1600, the forward gear ratios are provided by the following gears in mesh:

First $\dfrac{29 \times 39}{20 \times 15} = 3.77{:}1$

Second $\dfrac{29 \times 31}{20 \times 21} = 2.141{:}1$

Third $\dfrac{29 \times 24}{20 \times 25} = 1.39{:}1$

Fourth Direct drive

What alterations can be made to the ideal gear ratios, if they do not coincide with the gear wheel centres of the gearbox?

The ratio of one or more gear trains may be altered slightly. This will necessitate changes in number of teeth on the wheels, with corresponding changes in gear wheel centres.

How can the gear wheel centres be determined?

The distance from centre to centre of two gear wheels in mesh is the sum of their pitch circle radii (Fig. 11).

Let M = Module, i.e. the pitch diameter of the wheel in millimetres, divided by the number of teeth on the wheel

n = Number of teeth on the driver

N = Number of teeth on the follower

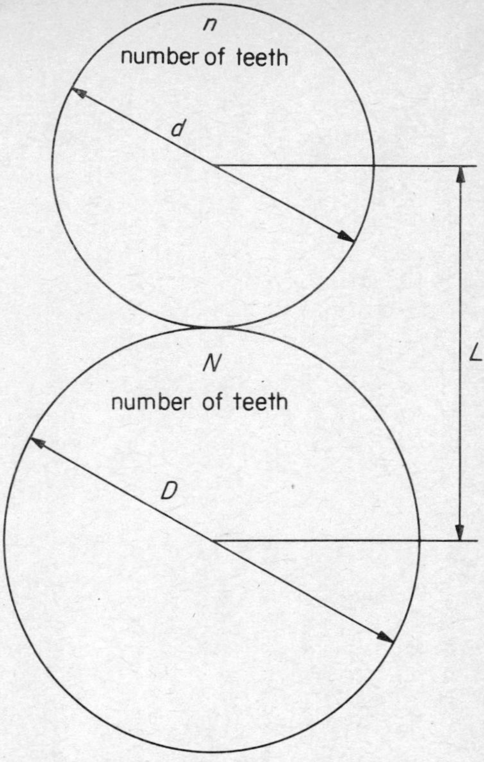

Fig. 11. Gear wheel centres

d = Pitch diameter of driver in millimetres
D = Pitch diameter of follower in millimetres
L = Distance between gear centres in millimetres

Then the gear ratio $= \dfrac{N}{n} = \dfrac{D}{d}$

Gear wheel centres $L = \frac{1}{2}(D + d) = (N + n)\dfrac{M}{2}$ mm

When several pairs of wheels of the same module are mounted on parallel shafts at fixed centres, then the sum of the teeth on each pair of wheels must give the same result. To accommodate certain ratios it may be necessary to alter the module on one or more pairs of wheels.

How can the number of teeth be altered in relation to the gear ratio?

Once the gear ratio and the shaft centres are known, the basic wheel sizes, or pitch circles, can be drawn, as in Fig. 11. The two discs making frictional contact would provide the same ratio, but under conditions of increased torque, slipping would take place. Toothed gearing does not slip, so that transmission of power is assured.

To find the number of teeth for a given module in terms of gear wheel centres, if the gear ratio $(r) = \frac{N}{n}$, the number of teeth on the larger wheel, or follower $N = rn$.

From

$$N + n = \frac{2L}{M} = rn + n = n\,(r + 1)$$

hence number of teeth on large wheel

$$N = \frac{2Lr}{M} \div (r + 1)$$

On the small wheel

$$n = \frac{2L}{M} \div (r + 1)$$

Example: Two gear wheels with teeth $5M$ are to work at 185 mm centres. Their gear ratio to be about 1.5:1.

To find number of teeth on wheels

$$N + n = \frac{2L}{M} = \frac{2 \times 185}{5} = 74$$

Number of teeth on small wheel

$$n = \frac{2L}{M} \div (r + 1) = \frac{74}{1.5 + 1}$$

$$= 29.6\text{, say 30 teeth}$$

Number of teeth on large wheel $N = 74 - n = 74 - 30 = 44$

The corresponding gear ratio will be $\frac{44}{30} = 1.466$, not 1.5

To obtain a ratio of 1.5:1, the gear centres could be reduced to 175 mm. Then $\frac{2L}{M} = \frac{2 \times 175}{5} = 70.$

$$n = \frac{2L}{M} \div (r + 1) = \frac{70}{2.5} = 28 \text{ teeth}$$

$$N = 70 - 28 = 42 \text{ teeth}$$

$$\text{Gear ratio} = \frac{42}{28} = 1.5{:}1$$

This illustrates the adoption of gear ratios, other than the ideal ratios, as calculated.

How are the gear teeth made to 'fit' or mesh with one another?

The pitch circles and the number of teeth on the wheels have been determined. For accurate machining dimensions for manufacturing, the designers would resort to gear tables for the required tooling information. For a general understanding of the design, consider the teeth shown in Fig. 12. They are positioned about the pitch circle, so that the distance from this circle to the tip of the tooth is less than the distance to the base (or root) of the tooth. A circle drawn through the tooth tips is called the addendum circle (a dimension added above the pitch circle diameter). A circle drawn through the base (or roots) of the teeth is called the dedendum circle (a dimension deducted from the pitch circle diameter).

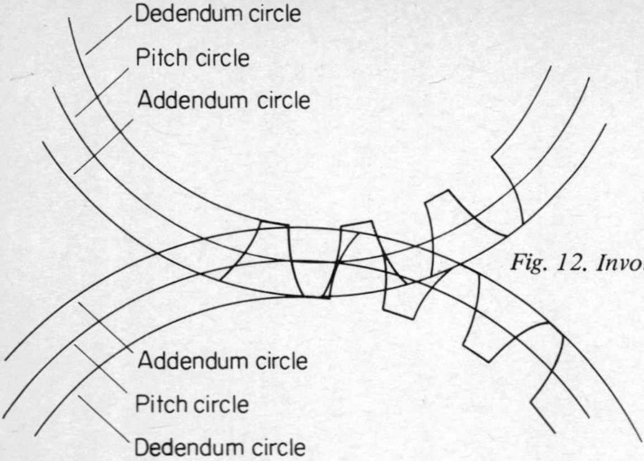

Fig. 12. Involute gears

The difference in depth, pitch circle to addendum, and pitch circle to dedendum, allows for clearance between the tips and the roots of the teeth when in motion.

The 'flank' or face of the tooth is based on the 'involute' curve. This can be generated by rolling a straight edge on the circumference of a circle, so that no slipping takes place. The involute is the path described by the end of the straight edge as it rolls about the circle (see Fig. 13). This design enables the tooth surfaces to roll on one another as power is transmitted.

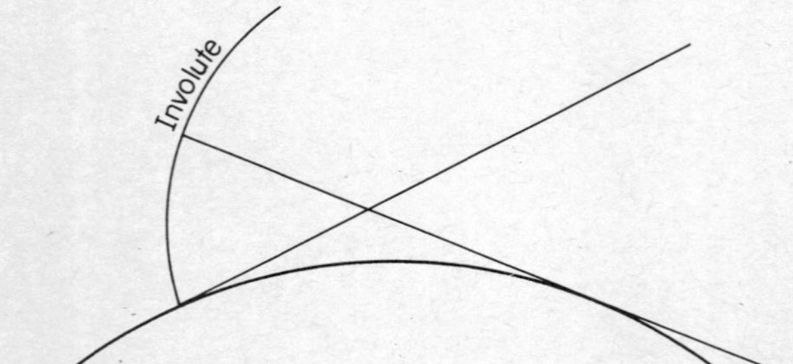

Fig. 13. Involute curve

What type of gearing can be used to transmit power at right angles?

Bevel gearing is used for the direct transmission of power between shafts whose axes intersect. They can be designed to work at any angle up to 180°, but 90° is most commonly used. (See final reduction gearing, page 129, and differential gears, page 133.)

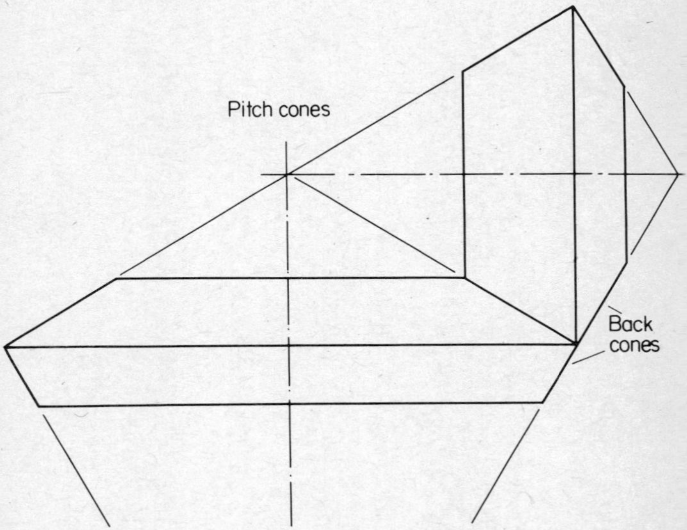

Fig. 14. Pitch cones and back cones for bevel gears

In this type of gearing the pitch circle diameters are set at right angles, so that the shaft axes intersect. The pitch cones are indicated by the lines from the pitch circle diameter to the point of intersection of the shafts. To construct gear teeth on these pitch cones, back, or normal cones are required, as shown in Fig. 14.

How are the gear teeth generated on this type of wheel?

From the apex of the back cone, with radius to the pitch diameter, the pitch circle of a spur gear forms the basis of a normal gear wheel design, as shown in Fig. 12. Returning to the bevel wheel (Fig. 15), the addendum and dedendum circles determine the face and root points of the tooth.

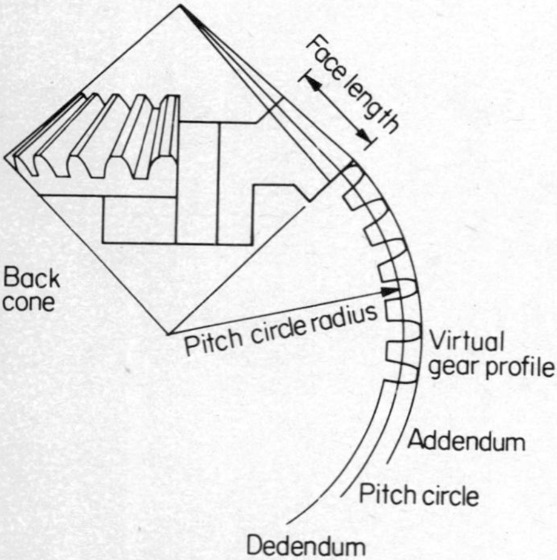

Fig. 15. Bevel gear design

The generated teeth are then wrapped around the back cone to further the formation of the teeth. The complete tooth is formed on lines passing back to the centre of the pitch cone and cut off to the required face length. The generated gear teeth would form a gear known as the 'virtual' gear wheel if the construction had been completed.

What is the relationship between the speed of shafts and the torque transmitted?

In Fig. 16 the gear reduction is 2:1 so that the shaft B rotates at 100 rev/min. If the torque on shaft A is 45 N m then the force between the gear teeth will be $\frac{45 \times 1000}{20}$ N. The torque on shaft B will be equal to $\frac{45 \times 1000 \times 40}{20 \times 1000}$ Nm. Thus when the speed is reduced by half the torque on the second shaft is doubled.

This also applies to bevel, and any other gearing in mesh.

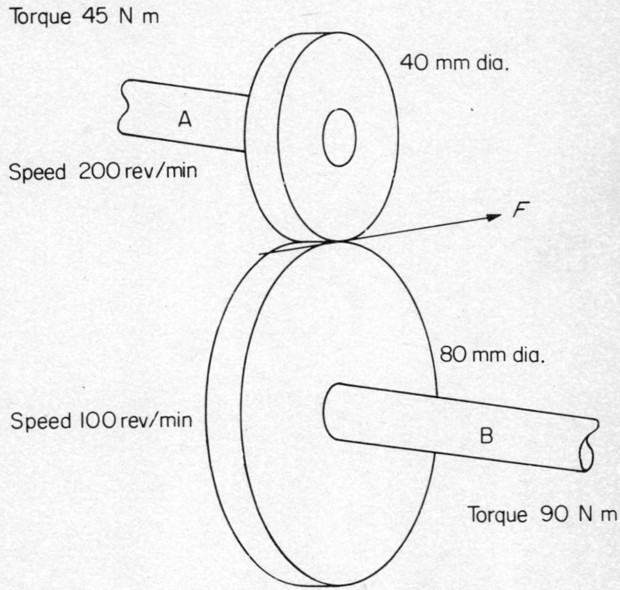

Fig. 16. Torque and speed of shafts

3

CLUTCHES

What is a clutch?

A clutch is a mechanism which enables the rotation of one shaft to be transmitted, at will, to a second shaft running on the same axis. Clutches may be divided into two distinct types: (a) positive clutches and (b) gradual engagement clutches.

In type (a) the clutch is either 'in' so that there is a rigid connection between the two shafts, causing them to revolve at the same speed, or it is 'out', when the shafts are disconnected. This type of clutch is not suitable for use between the engine and gearbox, but it is used in the gearbox and elsewhere in the transmission system.

In type (b) it is possible for one shaft (or the engine) to be revolving at high speed, while the other (or gearbox shaft) may be stationary or revolving at a lower speed. When the clutch comes into engagement the speeds of the two shafts will gradually become the same, and when it is fully engaged the shafts will revolve as one. In transmission systems with manual change gearboxes, the gradual engagement clutch depends for its action on frictional forces set up between two members being pressed together, as in gripping an item in a vice.

Why is some form of gradual engagement clutch required in a transmission system?

The need for a clutch arises from the fact that the internal combustion engine cannot start up against a load. If, for example, the engine of a motor vehicle is allowed to idle, and on declutching first gear is engaged, then without increasing the engine speed the pressure is released from the clutch pedal, the engine will stall. Therefore the drive must be disconnected and the engine speed increased (to produce a torque in excess of that required to move the vehicle), leaving the clutch to take up the drive on reconnecting it to the engine.

What are the basic principles applied to this form of clutch design?

Three distinct methods can be identified: friction, magnetic, and hydraulic.

Friction clutches may be constructed as a cone, a single-plate, or a multi-plate arrangement. Single-plate clutches are the most popular on private cars, almost eliminating the cone and multi-plate design as a transmission member immediately behind the engine. Small cone clutches are used in gearbox synchromesh devices (see Chapter 5), also in some overdrives (see Chapter 7). The multi-plate clutch (running in oil) is an essential feature of most automatic transmissions (see Chapter 9).

Magnetic clutches have been used from time to time, in conjunction with automatic or semi-automatic transmissions, to form a 'two-pedal' control system.

Hydraulic clutches, using fluid as the transmission material, appear as fluid flywheels and torque converters (see Chapter 4).

What are friction clutches?

Friction clutches rely on the nature of the contacting surfaces to produce a grip, assisted by some form of loading by means of coil springs or a diaphragm spring. The surfaces in contact are usually the steel or iron of the flywheel and the pressure plate and some friction material, such as Ferodo riveted to a centre plate or plates, mounted on the drive shaft to the gearbox.

Some multi-plate clutches use dissimilar metal, i.e. steel and phosphor bronze, while some automatic transmission clutch plates are lined with paper to obtain the desired characteristics.

What is the basic principle of friction clutch design?

The clutch must be capable of producing a frictional torque equal to, or greater than, the maximum engine torque.

Consider the single-plate clutch, Fig. 17, employing two friction discs; the total spring load W, multiplied by the coefficient of friction μ, will give the resultant tangential force F:

$$F = \mu W$$

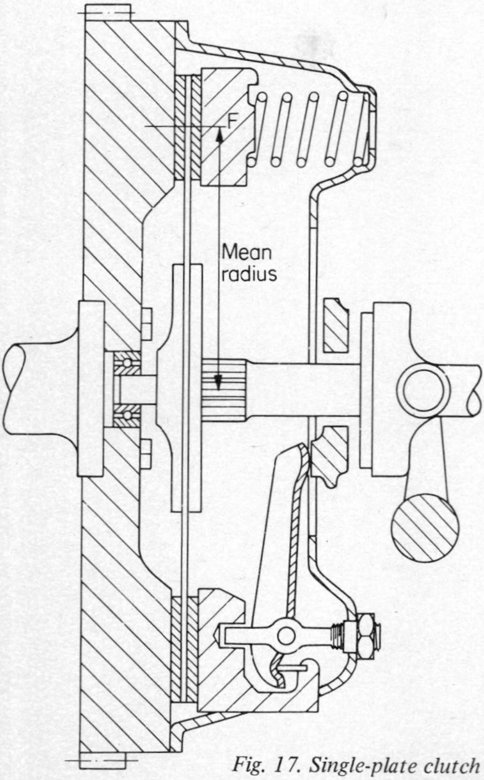

Fig. 17. Single-plate clutch

The torque transmitted will be equal to this force multiplied by a radius to the shaft centre. For an annular lining the effective radius can be taken as the arithmetic mean of the two radii. Thus

$$\frac{r_1 + r_2}{2} = \text{Effective radius}$$

The torque transmitted by one face would equal

$$T_f = \mu\, W\ \frac{r_1 + r_2}{2}$$

In a single-plate clutch there are two faces, but if n equals the number of faces the torque transmitted formula becomes

$$T_f = \mu\, W\ \frac{r_1 + r_2}{2} \times n$$

$$= \tfrac{1}{2}\,\mu W n\,(r_1 + r_2)$$

which may be used for any plate clutch calculations.

T_f	=	Frictional torque in newton metres (N m)
W	=	Total spring force in newtons
r_1 and r_2	=	Plate radii expressed in metres
n	=	Number of faces in contact
μ	=	Coefficient of friction between the surfaces.

Example: To find the maximum torque transmitted by a single-plate clutch with friction faces 0.3 m outside and 0.2 m inside diameters. There are ten coil springs applying a load of 50 kg.

Take $\mu = 0.3$

Total spring force $W = 10 \times 50 \times 9.81$

$= 4905$ N

$$r_1 = \frac{0.3}{2}\text{m and } r_2 = \frac{0.2}{2}\text{m}$$

There are two friction faces in contact, $n = 2$

$$\text{Maximum torque} = \tfrac{1}{2}\,\mu W n\,(r_1 + r_2)$$

$$= \tfrac{1}{2}\,0.3 \times 4905 \times 2 \;\; \frac{0.3}{2} + \frac{0.2}{2}$$

$$= \frac{0.3 \times 4905 \times 2 \times 0.5}{2 \times 2}$$

$$= 376.8 \text{ N m}$$

How are these design factors taken into account in the three types of friction clutch?

The single-plate clutch has a relatively small friction surface area, employing a clutch material having a coefficient of friction value from 0.3 to 0.45, the spring loading being supplied by a number of coil springs, or one diaphragm spring. The loading is relatively high.

The multi-plate clutch (Fig. 18), usually running in oil, will have a low coefficient of friction value of about 0.08 for steel on bronze (lubricated). This is compensated for by the in-

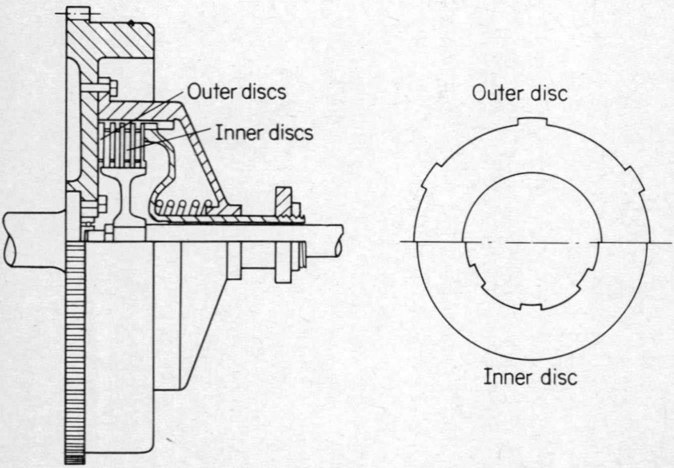

Fig. 18. Multi-plate clutch

creased area of contact, and by the larger number of working surfaces. The total spring load is then comparable with the single-plate clutch.

The cone clutch (Fig. 19) possesses a wedging action between the surfaces, so that a single, fairly strong spring can provide the required load. The single friction face, in this design, is about the same as that for one face of a single-plate clutch. The friction material will again be in the region of 0.3 – 0.45. Due to the wedging action, this clutch is fierce in operation.

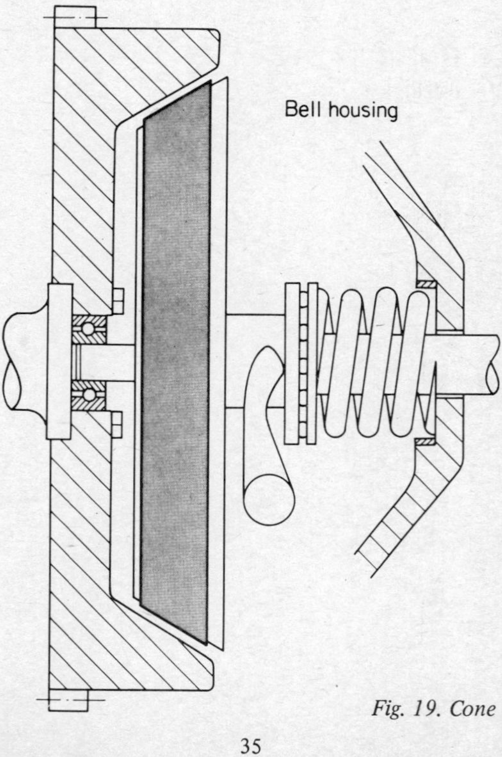

Fig. 19. Cone clutch

What advantage has a cone clutch over a plate clutch?

From the aspect of torque transmission, the cone clutch can transmit a higher torque for the same clamping force. This is achieved by the wedging action between the surfaces in contact. Fig. 20 shows an interesting example of a double-acting cone clutch, as used in the Laycock overdrive (see page 83). The friction lining on each face is about 15 mm wide, with a mean diameter of 140 mm, which appears to be a very small surface area for the torque transmitted.

How does the cone centre increase the force on the lining?

The action of the cone is similar to that of a wedge, where a comparatively small force exerted at the thick end produces a large force normal to each face. The theoretical value of this

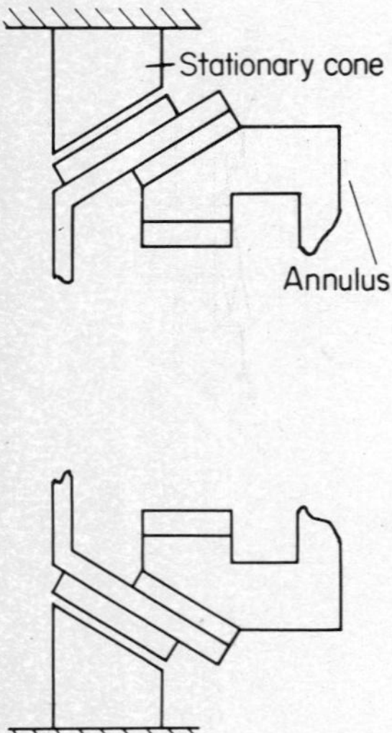

Fig. 20. Double-acting cone clutch (Laycock)

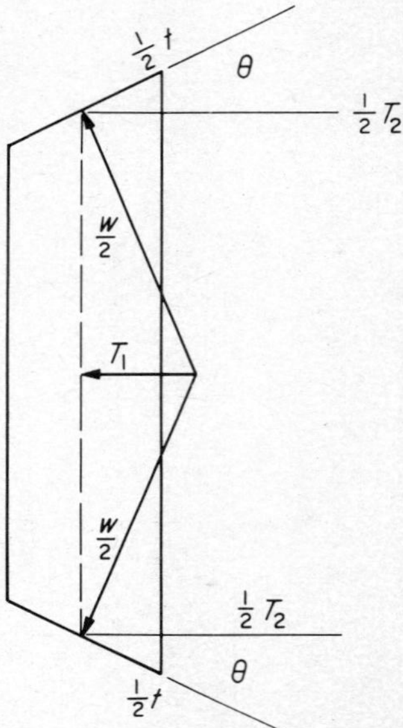

Fig. 21. Forces in cone clutch

force can be found from the triangle of forces. An additional force, opposing the motion of the wedge, is that of friction. The force applied to the wedge must overcome the frictional resistance before it can produce an increasing normal force.

Applying this reasoning to the cone (Fig. 21), T_1 is the axial thrust causing a force W normal to the friction surfaces. T_2 is the additional axial thrust to overcome the friction t to allow the clutch cone to engage fully. The clutch spring must supply sufficient force to overcome $T_1 + T_2$

θ = Half-cone angle.
μ = Coefficient of friction between lining and clutch face.

From the above, the total axial force $T = T_1 + T_2$.

$$\tfrac{1}{2}T_1 = \tfrac{1}{2}W \sin\theta, \text{ therefore } T_1 = W \sin\theta$$
$$\tfrac{1}{2}t = \tfrac{1}{2}W \text{ and } t = W$$
$$\tfrac{1}{2}t = \frac{\tfrac{1}{2}T_2}{\cos\theta} \text{ and } t = \frac{T_2}{\cos\theta} = W\cos\theta$$
$$T = W\sin\theta + W\cos\theta$$
$$= W(\sin\theta + \cos\theta)$$

Example: For a spring load of 150 kg, calculate the force normal to the clutch lining, if the included angle of the clutch is 25° and $\mu = 0.3$.

$$\text{Spring force in newtons} = 150 \times 9.81$$
$$= 1471 \text{ N}$$
$$T = W(\sin\theta + \cos\theta)$$
$$W = \frac{T}{\sin\theta + \cos\theta}$$
$$= \frac{1471}{\sin 12\tfrac{1}{2} + \cos 12\tfrac{1}{2}}$$
$$= \frac{1471}{0.2164 + 0.2929}$$
$$= 2888 \text{ N}$$

If the mean radius of the clutch lining is known, then the frictional torque can be found.

$$T_f = \mu W R$$

when μ = Coefficient of friction
W = Total force on lining
R = Mean radius of clutch face in metres.

What is the main advantage of the single-plate clutch?

The centre plate is light in weight and can accelerate and decelerate quickly during clutch operation, when the plate is released. (The inertia of a revolving object increases in direct proportion to increase in weight.)

Why is the centre plate so complicated in design?

The main object is to provide a flexible plate, that will take up the drive smoothly (Fig. 22). The hub is a separate part, sandwiched between the plates which make up the assembly, located initially by the coil springs in the boss.

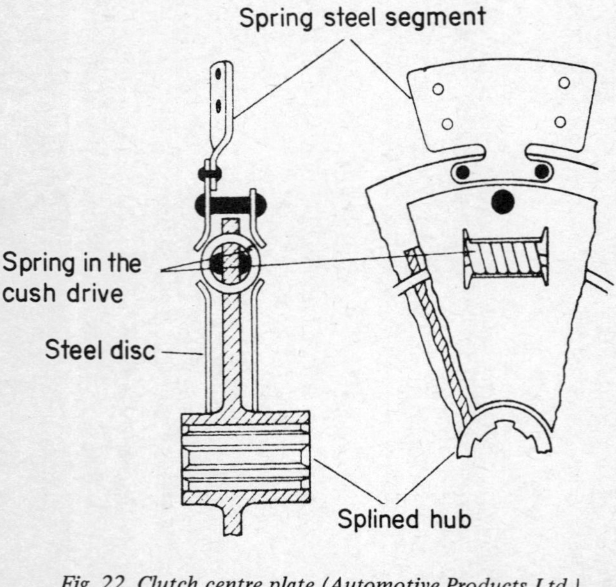

Fig. 22. Clutch centre plate (Automotive Products Ltd.)

Some movement is possible between the hub and the plate to form what is known as a 'cush' drive. This tends to damp out rotational 'snatch' on engagement of the clutch. The outer edge of the plate, supporting the linings, is segmented for flexibility. Each segment is also curved along its length to act as a leaf spring between the liners. This construction removes any fierceness that would arise if a continuous plate were used.

Is this type of centre plate construction used in all single-plate clutches?

No. Some of the power units adapted for front- or rear-wheel drive, with the clutch running in the gear train from the engine to the gearbox below, favour a solid steel disc equipped with the necessary linings and suitable mounting boss, without provision for any form of 'cush' drive.

Why does the pressure plate side of the centre plate generally wear quicker than the flywheel side?

Friction discs wear when there is slip, i.e. when there is relative movement between the flywheel and pressure plate with the clutch partially engaged. The disc contacting the pressure plate wears at a greater rate than the one contacting the flywheel, because the flywheel contains more metal than the pressure plate, providing a heat soak. This keeps down the operating temperature, reducing the rate of wear.

How is the clutch operated?

Both coil and diaphragm spring clutches are disengaged when the pressure plate is moved away from the flywheel. This is achieved by a release mechanism operated by the driver pressing the clutch pedal.

What types of release mechanisms are used?

Two are in general usage, mechanical and hydraulic. Fig. 23 shows a mechanical system consisting of a clutch pedal, suitable linkage, the release fork and the release bearing. Pressure on the clutch pedal causes the clutch release bearing to be pushed forward.

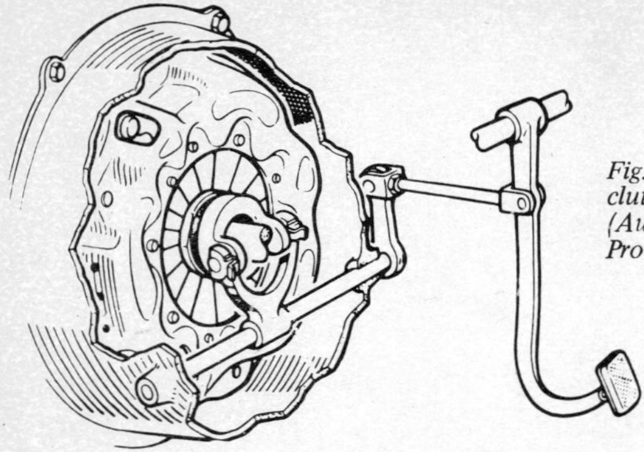

Fig. 23. Mechanical clutch control (Automotive Products Ltd.)

Fig. 24 illustrates the hydraulic system in which the connection between the clutch pedal and the release fork is achieved by the introduction of a master cylinder, flexible hose and a slave cylinder. This enables the power unit to be flexibly mounted without loss of effective action between pedal and release bearing.

What is the latest type of single-plate friction clutch?

The diaphragm spring clutch has superseded the coil spring clutch and is now fitted to practically all cars fitted with

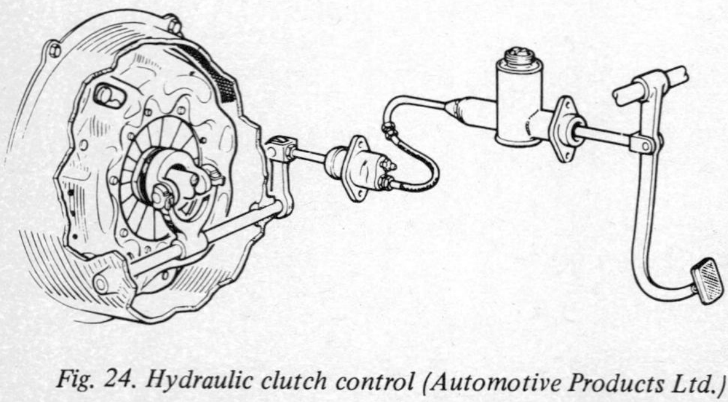

Fig. 24. Hydraulic clutch control (Automotive Products Ltd.)

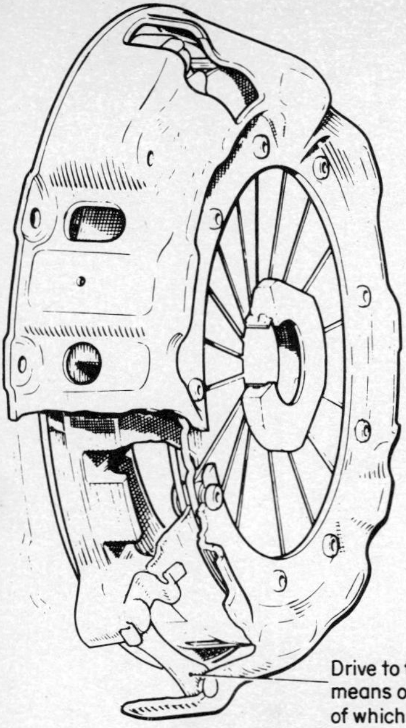

Fig. 25. Strap-drive diaphragm spring clutch (Automotive Products Ltd.)

manual gearboxes. Fig. 25 shows a strap-drive diaphragm spring clutch. The name 'strap-drive' arises from the fact that the pressure plate is attached to the cover plate by a set of three straps, so that it revolves as an integral part, at the same speed as the flywheel. In an alternative design, the drive to the pressure plate is transmitted by means of three lugs, similar to those used in the coil spring clutch (Fig. 17).

What are the advantages of the diaphragm spring clutch over the coil spring type?

(1) Less effort is required to release the clutch and hold it in that position. (Coil springs increase the reactive force as they are compressed.)

(2) The clamping force of the clutch mechanism increases as the clutch plate wears. (Coil springs decrease in their loading as wear on the plate causes them to extend.)

(3) Better balance is possible so that they can be fitted to engines capable of higher speeds.

(4) The compact design enables a shorter clutch housing to be used, hence a reduction in overall length of the power unit is possible.

(5) Elimination of levers, with associated pivot points and sliding surfaces, avoids squeaks and rattles due to wear.

Can friction clutches be controlled in any other way?

Yes. In certain automatic and semi-automatic transmissions the clutch pressure plate is actuated by centrifugal weights (Fig. 26). The pressure plate is held clear of the friction disc by light springs at idling speed, when the centrifugal weights

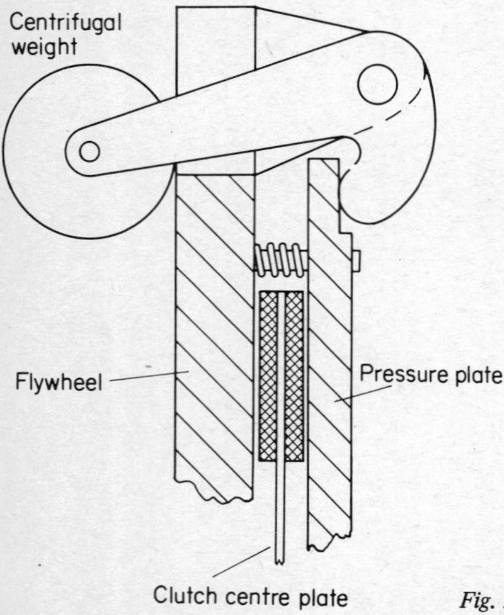

Fig. 26. Centrifugal clutch

are innermost, towards the clutch shaft. As the speed of the engine increases the weights fly outwards to operate the clutch levers, applying load to the pressure plate. Increase of engine revolutions per minute gives a corresponding increase in force on the friction lining.

Why are multi-plate clutches used in automatic transmissions?

The effective working diameter of the clutch has to be much smaller than the engine flywheel, so that by using a number of thin plates connected alternately to input and output shafts by suitable splines or serrations it is possible to achieve a large friction surface in a comparatively small space, see Fig. 27.

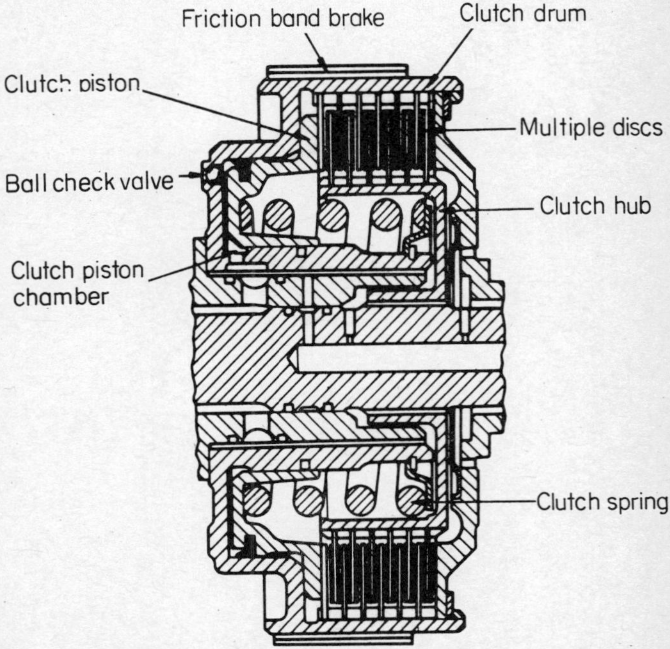

Fig. 27. Multi-plate clutch (automatic transmission)

The power transmitted can be calculated by using the formula developed for the single-plate clutch (page 33). The coefficient of friction between the plates will depend on the type of lining used and the fact that the clutch is running in oil.

How is this type of clutch activated?

By hydraulic pressure applied to the piston chamber, so that the piston moves to the right (Fig. 27), loading the clutch plates to lock the clutch hub to the clutch drum. In this manner two rotating parts of the transmission can be made to rotate as one.

What is the purpose of the check valve in the clutch housing?

When the apply pressure is released, the piston returning to the left (Fig. 27) by the action of the clutch spring pushes out the fluid through the return orifice. If the clutch is speeded up immediately after this release, oil remaining in the chamber will move outwards by centrifugal force, tending to move the piston to cause drag on the clutch plates. The check valve allows this residual oil to escape. At the next application of the clutch, normal pressure will seal the check valve orifice.

Is there an alternative to disc type clutches?

Yes. The Daf clutch, Fig. 28, has two sets of friction shoes attached to the flywheel. These shoes rotate inside a cylindrical drum, which drives the gearbox shaft. The construction is similar to a normal brake drum.

As the engine speed rises the first set of friction shoes flies out against the return spring load to contact the driving drum, exerting a force on its surface. The friction shoes develop a tangential force on the drum, producing a torque, which increases with engine speed. As soon as the driving torque overcomes the resistance of the transmission the drum starts to rotate. With further increase in speed the second set of shoes comes into contact with the drum, so that full torque can be transmitted.

This type of clutch is not suitable for use in conjunction

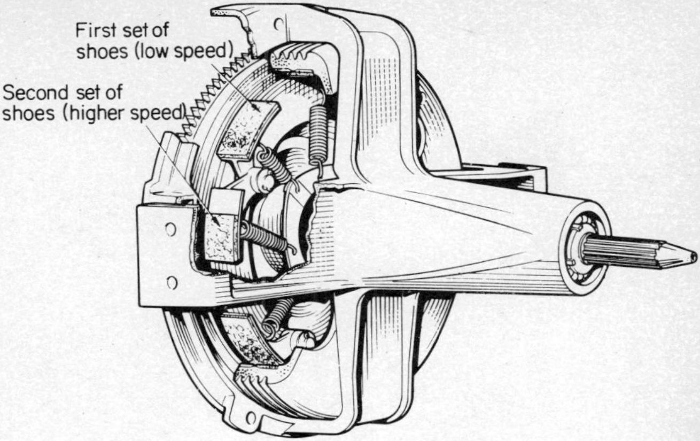

Fig. 28. Centrifugal clutch (Daf)

with a manual gearbox, as once the vehicle is in motion it is not easy to disconnect the drive, owing to the centrifugal action.

No such gear changing is required on the Daf car, hence the application as an automatic clutch. For operation of the transmission system, see page 115.

Can a clutch be controlled by an electromagnet?

The Ferlec clutch (Fig. 29), developed by Ferodo's French Company SAF, is an automatic clutch, effecting gear changes in the transmission system by means of a switch on the gear lever, which disengages or engages the clutch electrically.

The electromagnetic unit is mounted within a rectangular recess in the flywheel. Over the rear face of this coil is an armature ring. This ring is held against rotation, relative to the flywheel, by three trailing links at 120° to one another. Three compression springs are fitted in order to move the ring away from the coil when the current is switched off. A small clearance exists between the armature and the coil when the clutch is released. In addition to the armature plate there is an intermediate plate, or ring and a friction plate. Outside this there is a pressure plate which is attached by bolts to the armature plate. Outside this assembly are the slip rings and the brush holders.

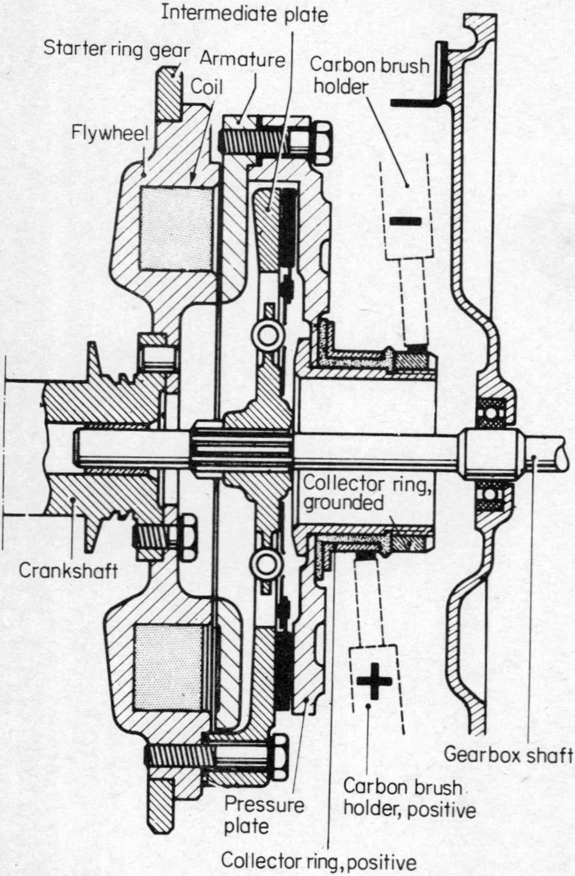

Fig. 29. Ferlec magnetic clutch (Renault)

When the current to the coil is switched on, the armature plate is attracted to the flywheel, taking with it the pressure plate which is bolted to it. Thus the pressure plate moves toward the intermediate plate to grip the friction plate and activate the clutch, as both the pressure plate and the intermediate plate rotate at engine speed.

The amount of pressure applied varies with the current passing, so that a progressive engagement can be obtained by increasing the current supplied to the coil. The supply from the vehicle's dynamo increases with engine speed as the car accelerates, to cause clutch engagement. When the accelerator pedal is released the clutch disengages because the electrical supply is reduced and the compression springs separate the armature from the coil.

Is there a magnetic clutch using powdered iron to transmit torque?

The original idea of this clutch was patented in the USA by J. Rabinow and developed by the Eaton Manufacturing Co. in that country. It has been made and developed in this country by S. Smith & Sons (England) Ltd., for use on semi- and fully automatic transmission systems.

The coupling consists of two main members (see Fig. 30), one attached to the flywheel, the other connected to the gearbox or transmission so that a small clearance, or gap, exists between the periphery of the inner (or gearbox) member and the inner surface of the flywheel member. This gap is filled with fine iron powder, which finds its way into the gap by centrifugal force when the engine starts. No power can be transmitted at this stage as there is no force to hold the powder in position. When a current is passed through the excitation coil, the magnetic flux created will cause the iron powder to act as a solid coupling, provided the current value is sufficient. At lower values there will be a certain amount of slip between the driving and the driven members, simulating the action of the friction clutch.

Current is supplied through the brushes and slip rings shown on the left of Fig. 30. The generator of the car is only

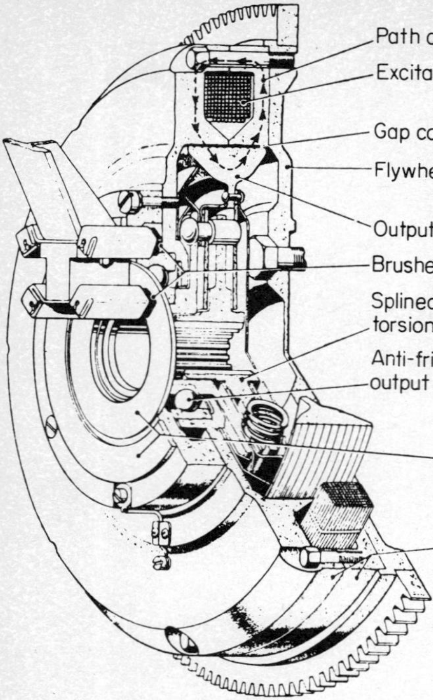

Fig. 30. Magnetic clutch (Smiths, Cricklewood)

employed as the source of current supply at relatively low road speeds. When the vehicle has accelerated to approximately 16 km/h a sensitive road speed switch restores the generator to its normal function and at the same time routes current directly from the battery to the coupling.

What is a dog clutch?

A dog clutch is a positive engagement clutch, in which the drive is connected or disconnected by sliding a selector along a splined shaft (Fig. 31). The pinion carrying one set of dogs revolves on a plain bush, while the selector is internally splined to mate with the shaft. On engaging the clutch, the pinion is locked directly to the shaft via the selector. Dog clutches are used in the majority of gearboxes, as they are incorporated into most synchromesh mechanisms.

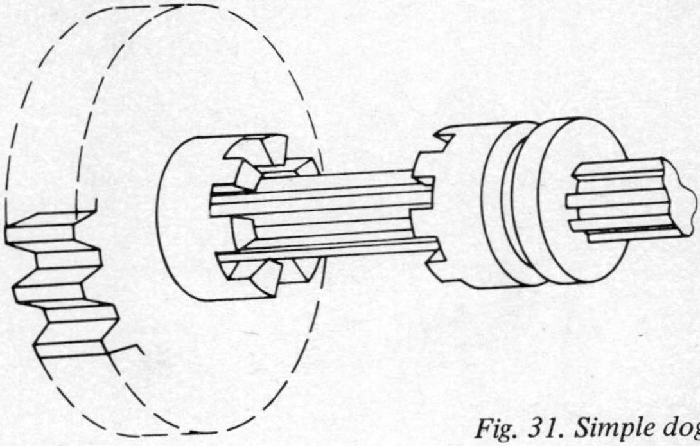

Fig. 31. Simple dog clutch

What is an overrunning clutch?

An overrunning clutch (or one-way clutch) permits motion to be transmitted through it in one direction only.

If two shafts A and B are connected by such a clutch, motion will be transmitted when, say, shaft A is turned clockwise. If shaft B is turned clockwise, this shaft will run on without imparting motion to shaft A. If the shaft A is turned anti-clockwise it will also overrun without imparting motion to shaft B. This type of clutch is used in many automatic transmission systems and overdrives.

In the roller clutch illustrated in Fig. 33 the outer member A may be considered to be attached to shaft A in the above discussion.

What types of uni-directional clutch are in general usage?

Sprag and roller clutches are the most popular.

The full-phased sprag clutch shown in the engaged and disengaged positions (Fig. 32 (a) and (b)) consists of an inner

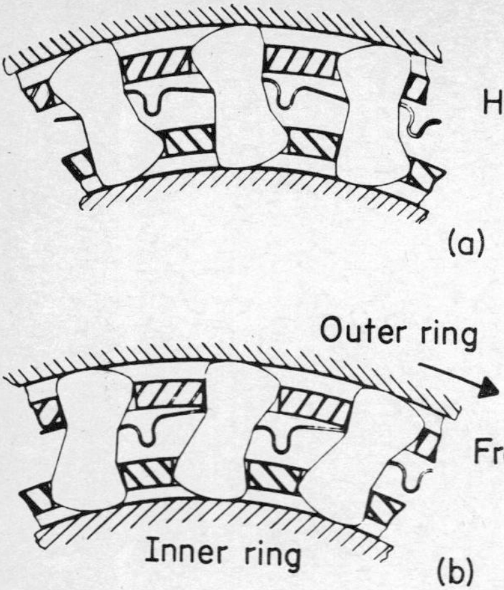

Fig. 32. Sprag clutch

and outer race, with control rings held concentric by their respective races, a number of sprags and a thin plate spring passing round the clutch to apply a slight load to the sprags.

When the outer race is running faster than the inner race in a clockwise direction, as at (b), the outer race rocks the sprags

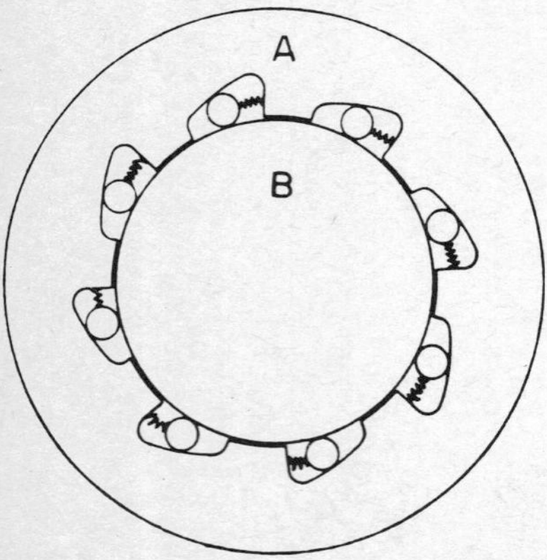

Fig. 33. Roller one-way clutch

over in the same direction, owing to the slight frictional contact between the surfaces, allowing the clutch to slip. When the inner race tends to run faster than the outer race, as at (a), in a clockwise direction, the clutch 'locks up', by the fact that the contact made with the lower surface of the sprags rocks them over, in the direction of motion, to act as struts between the inner and outer races.

The roller one-way clutch (Fig. 33) is simpler in design. Rotation of the inner race, clockwise, at a higher speed than the outer race gives a free-wheeling condition, as the rollers will tend to move against the spring load to increase clearance. When the outer race tends to run at a higher speed than the inner race (clockwise) the clutch will 'lock up', the outer race causing the rollers to wedge between the inner member and the ramps of the outer member. The outer member can also revolve anti-clockwise about the inner member without lock up occurring.

4

FLUID FLYWHEELS AND TORQUE CONVERTERS

How does the fluid flywheel operate?

When the engine is set in motion, the pump A (Fig. 34) rotates with the oil in its cells; centrifugal force is thus impressed on the fluid, causing it to move towards the outer circumference of the pump. Assuming that the turbine B is stationary (as when the car is at rest) the oil ejected from the pump causes the oil in the turbine to move toward the centre, feeding the pump intake, and a circulation is set up. This circulation is in the form of a helix passing round the guide ring, its pitch depending on the difference in speed between the pump and the turbine.

At starting, when the turbine is at rest, the pump jets will pass over a number of cells before entering the turbine; as the speed of the turbine increases the pitch of the helical flow will decrease until it is passing in and out of adjacent cell apertures. Generally, the turbine is running at a lower speed than the pump. When the difference in speed is great, a high vortex exists in the coupling; when the difference is small a low vortex exists. If the pump and turbine run at the same speed, say at a point when the car is about to overrun the engine, no vortex will exist for the time being until one member slows down. If at any time the turbine runs faster than the pump, it will throw fluid back into the pump and the flow will be reversed. This condition would exist when towing a car fitted with a fluid flywheel; the turbine would then turn the pump and the engine could be turned to effect a start.

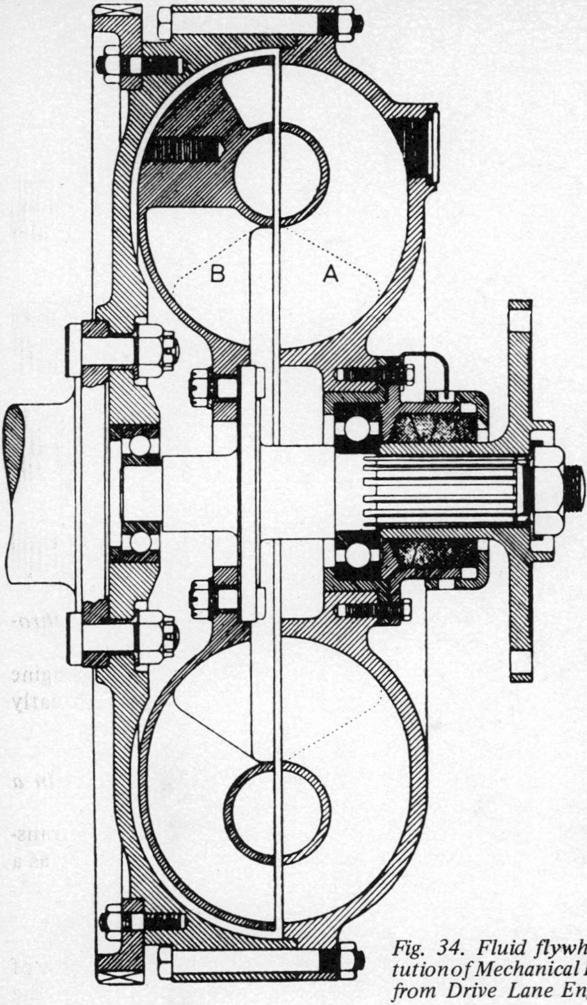

Fig. 34. Fluid flywheel (Institution of Mechanical Engineers, from Drive Lane Engineering)

What is the purpose of the fluid flywheel?

It replaces the clutch in some automatic and semi-automatic transmissions. It is in fact a hydraulic coupling, allowing 100% slip at low speeds, increasing to 98% grip under full load conditions without external control.

Why is the turbine B placed forward of the pump A?

The reason for this construction is to take care of the forces that tend to separate the two sections when the flywheel is in operation, by introducing thrust bearings on the turbine shaft.

Does the fluid flywheel multiply torque?

No. It simply transfers energy from pump to turbine in relation to engine speed. Generally the turbine is rotating at a slower speed than the pump, creating a helical flow of the fluid through the blades.

What servicing is required to maintain the fluid flywheel?

Periodic checks to see that it has the right amount of fluid are all that is necessary. Provided that no leakage occurs at the output shaft topping up should not be required.

Can the fluid flywheel be used in conjunction with a synchromesh gearbox?

No. There is a considerable drag present at low engine speed, so that a friction clutch must be added; as in some early Chrysler, etc., transmissions.

Can the fluid flywheel be used in any other position in a transmission layout?

Yes. In some early General Motors Hydra-matic transmissions, two couplings were used, the second one acting as a shifting coupling, i.e. to change ratio in the gearbox.

Can a vehicle fitted with a fluid flywheel be tow-started?

Yes, provided that the car is towed at a speed in excess of 48 km/h. The turbine will then act as a pump to turn the flywheel member and rotate the engine.

What is a torque converter?

In its simplest form a torque converter has three elements working in a closed circuit (see Fig. 35), consisting of (a) a pump rotated by the engine, (b) a turbine attached to the input shaft of the gearbox, and (c) a reactor (or stator) mounted on the gearbox casing by means of a one-way clutch. The unit has two phases of operation, first as a torque converter and then as a fluid flywheel.

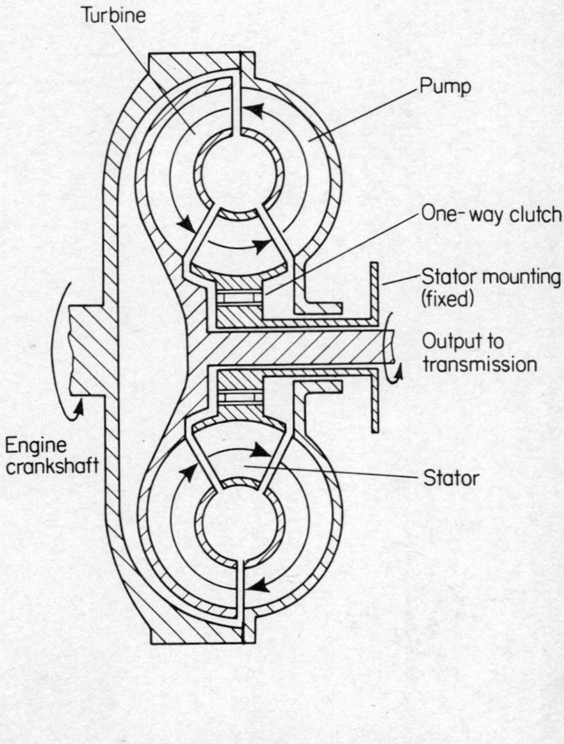

Fig. 35. Single-stage torque converter

How is the fluid flow arranged within the unit?

The pump is fitted with blades that are very nearly radial, and which on rotation increase the energy of the fluid, as in the case of the fluid flywheel. The fluid is now thrown into the turbine blades, which are generally curved, so that a comparatively small amount of energy is required to drive them. Owing to the curvature of the blades the discharge from this element is opposite to pump rotation and is corrected by the stator, which has a series of blades curved in the reverse direction to those on the turbine. In this way the fluid is re-directed to enter the pump in the direction of engine rotation.

What determines the two phases of operation?

The stages of operation are controlled by the direction of the fluid flow as it leaves the turbine blade (Fig. 36). This in

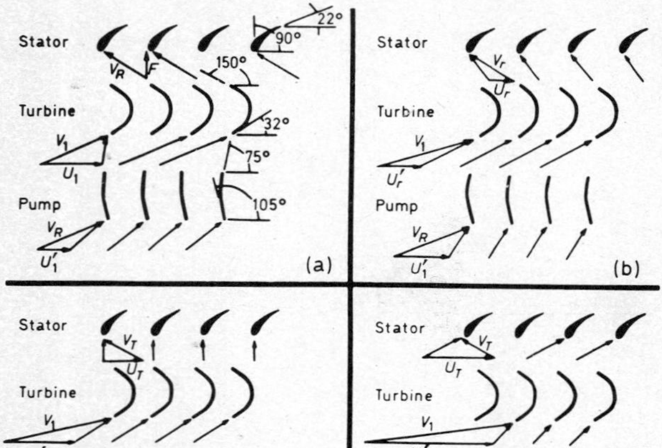

Vector diagrams at (a) stall; (b) 0.5 speed ratio; (c) 0.7 speed ratio; (d) coupling points

Fig. 36. Direction of flow between elements in single-stage converter (S.A.E. and V.J. Jandasek)

turn depends on the speed ratio that exists between the pump and the turbine. At low engine speeds (Fig. 36 (b)) the flow out from the turbine tends to turn the stator in the reverse direction to the engine, locking it by means of the one-way clutch. The stator remains locked until the turbine is running at about 70% of the engine rev/min (this will depend upon design features). This speed ratio is sometimes referred to as the coupling point, i.e. when the discharge from the turbine is striking the stator blades end-on (Fig. 36 (c)).

Beyond this speed ratio, the turbine discharge is in the same direction as the engine rotation (Fig. 36 (d)); this unlocks the stator, causing all three members to rotate with the engine. Torque multiplication ceases and the unit acts in the same manner as a fluid flywheel.

How is the torque increased?

The torque conversion takes place during the first stage when the stator is locked. The energy imparted to the fluid by the pump is only partially absorbed by the turbine, so that the fluid enters the stator still possessing a proportion of this energy. The stator re-directs the fluid into the pump without any further appreciable loss. Here the energy is increased by the power of the engine building up additional torque within the unit.

Can the torque multiplication stage be extended?

Multi-stage converters are created by increasing the number of turbines and/or stators. Fig. 37 shows a twin-turbine arrangement, with two stators or reactors, together with a built-in epicyclic gear. These designs reduce the number of required gear ratios, so that such a converter could provide all the transmission requirements by the simple addition of a forward and reverse drive unit.

A further refinement is that the second reactor is a two-position arrangement, allowing for the converter torque multiplication to be varied within the unit while the engine is running.

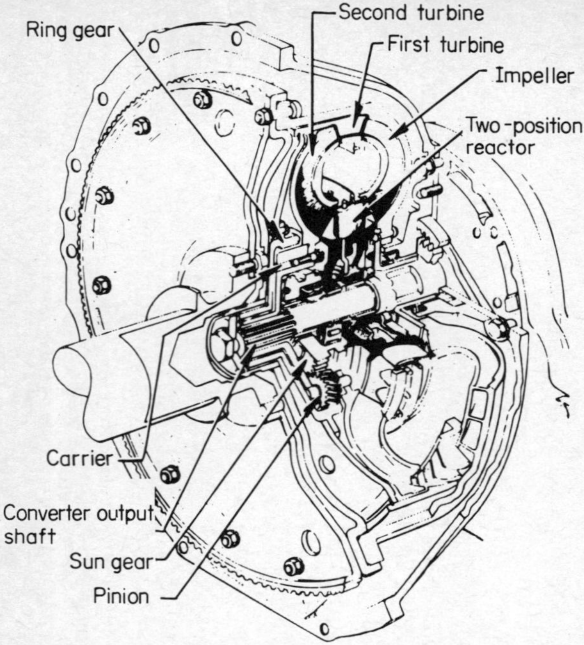

Fig. 37. Twin-turbine torque converter (General Motors)

What servicing does the converter require?

The converter is sealed by welding and is serviced by replacement only. No filler or drain plugs are fitted, as the transmission oil supply feeds the converter.

Can the converter be checked for serviceability?

A stall-torque test, in which the turbine is held stationary while the engine is driving the pump, will assess the condition of the converter.

The pump will set up a flow of fluid through the stationary turbine to lock up the stator on the one-way clutch. If the one-way clutch, between the stator and its mounting, is not functioning the stall torque speed will not be achieved. The reduction in speed is due to power absorption between the pump and the stationary turbine, similar to the brake power test on an engine.

It is essential that the engine is capable of developing full power before the test is carried out, as this will affect the stall-torque reading on the tachometer.

To carry out the test an accurate tachometer is required, as well as the stall speed of the engine, from the car manual. When the tachometer is connected and the engine is at running temperature, the hand and foot brake are applied, then with lock selected on the transmission drive selector the engine speed is increased by means of the accelerator pedal. The engine speed will 'settle' to a steady value, which must be read within ten seconds of increased engine speed. Compare the value indicated by the test against the value given in the manual.

If the stall speed is up to 300 rev/min below that specified, the engine is not developing full power. If it is more than 600 rev/min below that specified, the converter assembly must be replaced.

What other symptoms can indicate converter faults?

Inability to start on steep gradients, combined with poor acceleration from rest, indicates that the converter one-way clutch is slipping or that the stator support is fractured.

Below standard acceleration in third gear above 48 km/h combined with poor maximum speed, indicates that the stator one-way clutch is locked in the engaged position.

In these circumstances the stator will not be able to rotate with the pump and turbine, so that the fluid flywheel stage cannot occur.

5

GEARBOXES

What is a layshaft gearbox?

This type of gearbox is arranged as two parallel shafts (Fig. 38), with the drive entering from the engine on the primary or first-motion shaft, driving through the layshaft by constant mesh wheels and out to the mainshaft by the gearing in mesh. A fourth shaft carries the reverse gear wheel to change direction of rotation when this gear is selected. Sliding mesh, constant mesh and synchromesh are methods of gear selection used in layshaft type gearboxes.

What type of gearing is in general use?

Spur gearing is used for transmitting power between parallel shafts, see page 61. The gear teeth may be straight cut or helical in pattern.

What are straight-cut gears?

As the name implies, the teeth are cut straight across the edge of the wheel at right-angles to its face. The use of this type of gearing within the gearbox is confined to sliding mesh systems.

What is helical gearing?

This is a form of spur gearing in which the teeth are set at an angle to the face and to the axis of the wheel. This form of gearing is used in constant mesh and synchromesh gearboxes. The gearing cannot slide sideways into mesh, unless it is mounted on a shaft possessing the same helix angle.

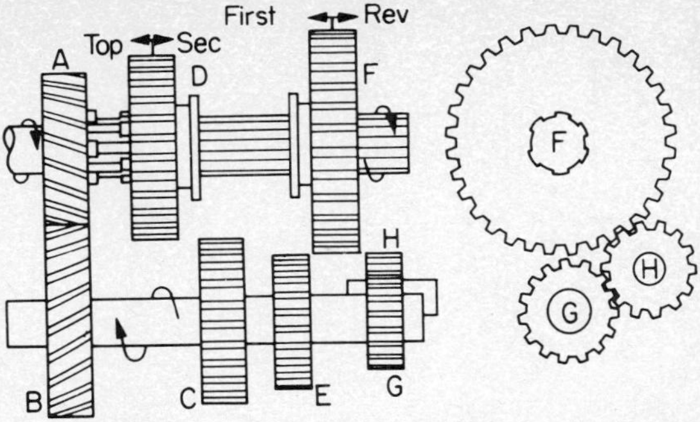

Fig. 38. Three-speed and reverse sliding mesh gearbox

What are gear trains?

A number of gear wheels in mesh to provide a speed reduction, or a speed multiplication, is called a gear train. Simple gear trains have all the gear wheels in the same plane. Compound trains, as in some automobile transmissions, consist of two or more simple trains in series.

What are 'idler' gears?

These are gear wheels introduced into the train with the sole purpose of reversing the direction of motion. Idler wheels do not alter overall gear ratios.

What is a sliding mesh or crash type gearbox?

A three-speed and reverse gearbox of the sliding mesh type (Fig. 38) is composed of four shafts. The first motion shaft extends from the engine flywheel, through the clutch centre plate to the pinion within the gearbox. The end of the first motion shaft houses a spigot bearing within the pinion to support the front end of the mainshaft. This shaft extends through the rear of the gearbox to the driving flange, connecting it to the universal joint. It is also splined throughout its length, so that the two pinions mounted on it (with mating splines) can slide sideways, but cannot rotate without imparting motion to the shaft. The layshaft may be positioned directly below the mainshaft, carrying four pinions as a cluster of gears, i.e. no pinion can revolve independently.

The reverse motion shaft is positioned so that the pinion meshes with the smallest layshaft wheel and the largest mainshaft wheel to form an idler gear in train, reversing the direction of motion of the mainshaft.

How is the power transmitted through the gearbox, in each gear?

Using a three-speed and reverse gearbox, as in Fig. 38. First gear is engaged by moving pinion F to the left, to mesh with pinion E on the layshaft. Power is transmitted from the first motion shaft to the layshaft from pinion A to B and out to the mainshaft from E to F. Second gear is engaged by moving the gear lever through the neutral position, to move pinion D to the right to engage with pinion C on the layshaft. Power will now pass from A to B as in the previous gear, but the route out of the gearbox will be via C and D. Top gear is engaged by moving pinion D to the left to engage the simple dog clutches between pinions A and D. Power is now transmitted direct from the first motion shaft to the mainshaft. The layshaft is still rotating by virtue of the connection from A to B.

Reverse gear is brought into action by moving the pinion F to the right, so that it meshes with the reverse (idler) pinion H. The drive will now pass from A to B and out to the mainshaft via G, H, and F.

What constitutes a constant mesh gearbox?

Any gearbox in which the pairs of gear wheels are always in mesh. Fig. 39 shows a three-speed and reverse gearbox of this type with simple dog clutches. In this arrangement an additional wheel is required on the mainshaft. It should be noted that the wheels on the mainshaft must be mounted on separate bearings, so that they can revolve independent of the shaft.

How are the gear ratios selected?

The gear selector lever operates the selector rods and the

forks to move the dog clutches on the mainshaft into mesh with the required pinion A, C, E or J.

First gear is selected by moving dog clutch 2 to the left to mesh with the dogs on pinion E. Power is transmitted through the box via pinions A, B, F and E. Second gear is selected by moving dog clutch 1 to the right to engage with the dogs on pinion C. Transmission of power in this ratio is via A, B, D and C. Top gear is selected by moving dog clutch 1 to the left. This action locks shaft A to the mainshaft via the dog clutch.

Reverse gear is obtained by moving dog clutch 2 to the right. Power is transmitted via A, B, G, H and J.

Note that when the engine is running the wheels C, E and J are revolving on the mainshaft, as they are in constant mesh with the layshaft and the first motion shaft, pinion J revolving in the opposite direction to C and E.

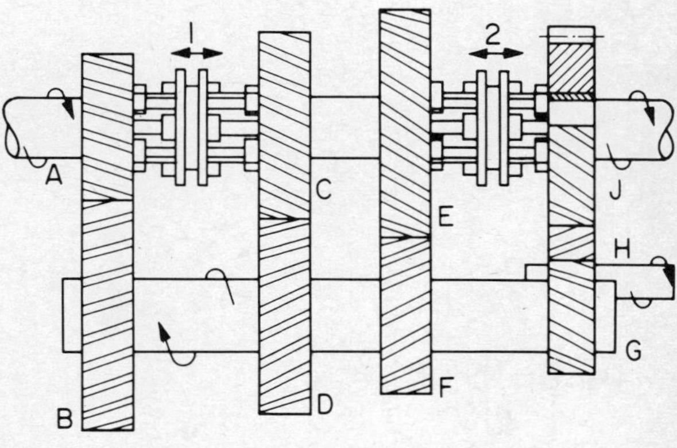

Fig. 39. Constant mesh gearbox

What does the term 'synchromesh' mean?

In any synchromesh system the speeds of the two shafts or gears to be engaged are synchronised before the change in gear ratio is actually effected.

Why is helical gearing used in constant mesh and synchromesh gearboxes?

As the pinions are always in mesh they can be made to closer limits. The helix angle also increases the length of the tooth, so that the tooth loading, for a given pinion width, is thereby reduced. In this form of gearing more than one tooth is in engagement at any one time, contributing to quieter operation.

Is there any disadvantage in using this form of gearing?

Yes. When a pair of wheels are in mesh and transmitting power, a side-thrust is developed owing to the helix angle. The layshaft wheels have the helix angle in the same direction (see Fig. 39), so that the thrust from one pair of wheels is opposed by the side-thrust from the second pair of wheels in the gear train. Appropriate thrust washers are fitted on the mainshaft immediately behind each pinion.

What constitutes a synchromesh device?

Any mechanism replacing the simple dog clutch which involves (or embodies) a friction device (such as a cone clutch) to bring about synchronisation of the shafts and wheels before the actual engagement of the sleeve to form a positive drive.

What opposes the side-thrust emanating from the shafts?

Retaining rings fitted to the bearings at the ends of the gearbox casing carry the thrust. The side-thrust arising from one pair of wheels may not be fully balanced by the end-thrust from the following pair of wheels, so that the resultant thrust is passed through the gearbox bearings to the casing via the thrust ring.

What are the advantages of synchromesh devices?

With the latest designs, gear changes can be made without any noise during the change of ratio. The time taken to change is much less than in a sliding mesh gearbox, resulting in greater distance covered in comparison.

What is a constant load synchromesh device?

In early synchromesh systems, the total load that could be applied to the friction clutch was limited by a system of ball locks positioned between the hub and the sleeve of the unit (Fig. 40).

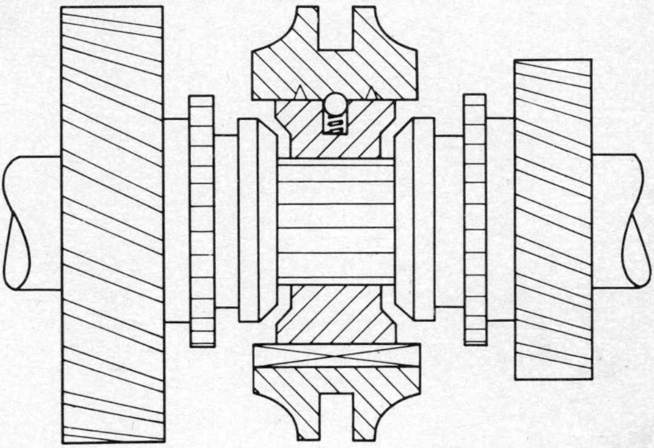

Fig. 40. Constant load synchromesh

How does the constant load device work?

On changing gear the synchro hub and sleeve (Fig. 40) move along the shaft as one unit, held together by the ball locks between their sliding surfaces. The first contact with the gear pinion is made by the conical surface of the hub encountering the conical surface of the gear. A clutch action follows, whereby the pinion and the shaft are made to revolve at the same speed. The outer sleeve now moves along the hub (depressing the ball locks) to make contact with the synchromesh teeth mounted on the pinion. The pinion is now connected to the mainshaft via the splines of the sleeve and hub respectively.

Why has the system become obsolete?

A smooth, silent gear change can only be achieved if sufficient time is given to bring about synchronisation. If a fast gear change is attempted the ball locks can be depressed before synchronisation takes place, allowing the splines of the sleeve to contact the synchro-dogs of the pinion 'end-on', creating noise and causing damage to the unit.

What systems have superseded the constant load device?

The inertia lock device, in which an increasing load could be applied to the cone clutch, was the initial advance on the constant load system. Baulk-ring devices are now the most popular, mainly due to economics in manufacture.

How does the inertia lock operate?

The brake-ring A (Fig. 41) has three equally spaced fingers pointing inwards toward the shaft centre. At the base of the fingers, and placed in grooves which pass through the selector B, are three special plate springs. These springs apply a slight pressure to the brake-ring, keeping it in contact with the conical surface of the gear pinion. As the pinion revolves, a slight drag between the surfaces causes the brake-ring to move,

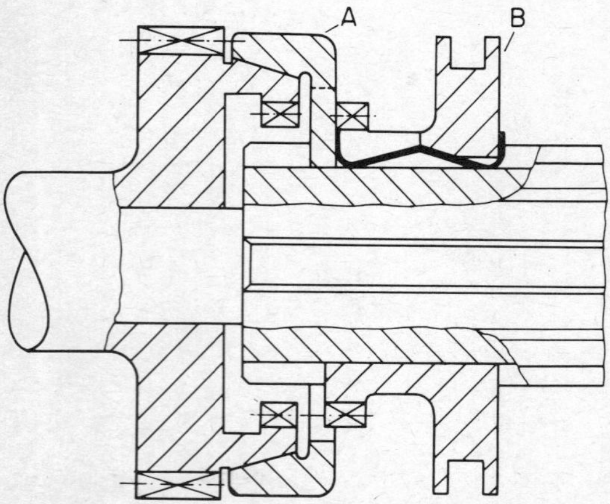

Fig. 41. Inertia lock system

so that the fingers are up against one side of the grooves in the mainshaft.

When the selector moves towards the pinion, the brake drag is increased by the deflection of the plate springs. Passing over the springs, the selector now makes contact with the brake-ring because of a slight overlap which exists between the brake fingers and the selector abutments. Any increase of pressure on the gear change lever means an increase of pressure on the brake surfaces, bringing about synchronisation in relation to the applied load.

At the point of synchronisation the brake fingers leave one side of the groove to pass over to the other side. In so doing the overlap between the fingers and the selector is reduced to zero, and the selector moves through the brake ring to engage with the internal splines of the pinion. The gear is now connected to the shaft via the splines of the selector.

Why is it called 'inertia lock'?

Because it relies on the inertia of the brake-ring for its operation. Consider the shaft (Fig. 42 (a)) rotating clockwise at a speed greater than that of the brake-ring. The inertia of the brake-ring will cause the fingers to assume the position shown. Should the speed of the brake-ring be higher than that of the mainshaft, the fingers will take up the position shown in Fig. 42 (b). This inertia effect prevents the selector passing along the shaft until synchronisation is achieved.

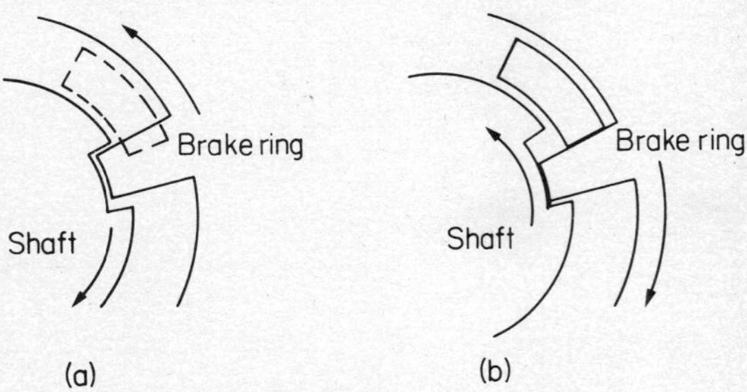

Fig. 42. Inertia effect

What is a 'baulk-ring' system?

The baulk-ring device (Fig. 43 (a), (b) and (c)) consists of a driving gear A on which there is a ring of clutch teeth B and a steel cone C. The sliding sleeve D has internal teeth in the bore; these engage with the splined interior of the hub member E, and when required, with the clutch teeth B. Between the hub member E and the driving gear A is located a baulk ring F, the bore of which is coned to fit the steel cone C. On the outside of the baulk ring are teeth G which correspond to the clutch teeth B. On the side adjacent to the sliding sleeve D these teeth are chamfered at a specially determined angle, as also are the internal teeth P on the sliding sleeve D.

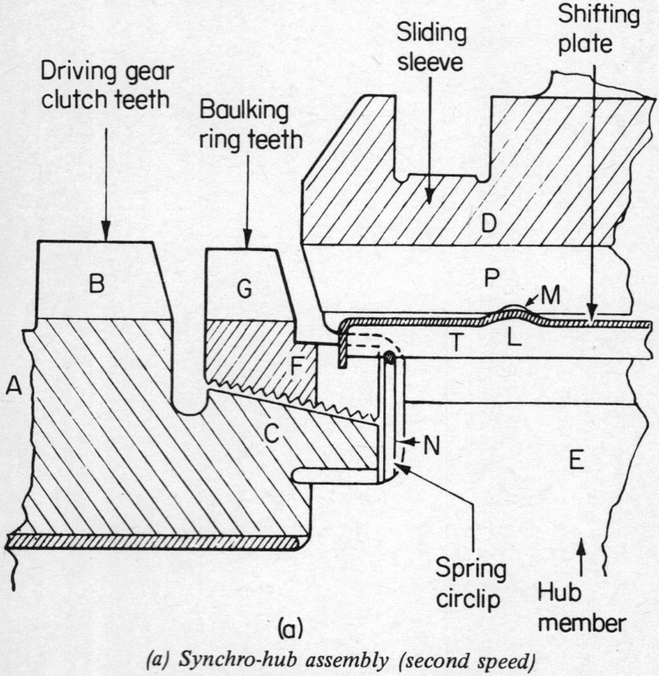

(a) Synchro-hub assembly (second speed)

Fig. 43. Baulk-ring synchromesh (Chrysler)

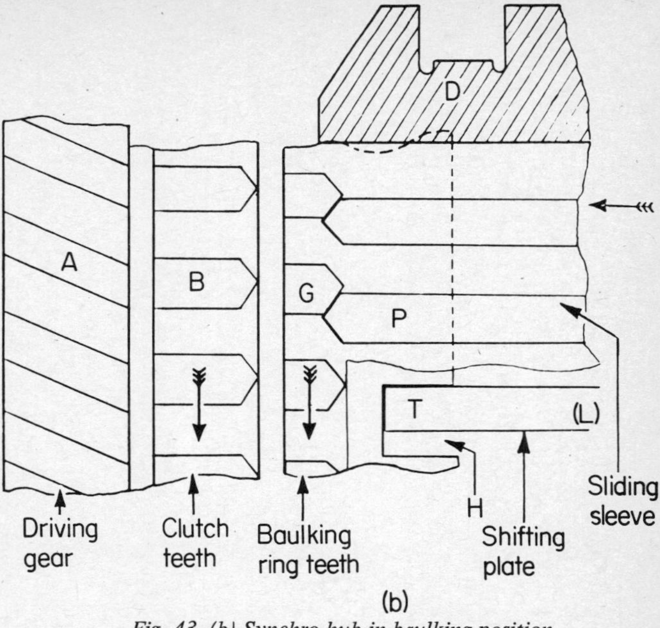

Fig. 43. (b) Synchro-hub in baulking position

The baulk ring has three slots H (Fig. 43 (b) and (c)) which engage with the ends of the shifting plates T. These plates are positioned in slots cut along the hub member, and they are of such width that the baulk ring is permitted to take a partial rotation in either direction. On the shifting plates there is a small protrusion L which locates in the groove M in the bore of the sliding sleeve at the neutral position. Spring circlips N load the shifting plates in an outward direction, so maintaining a slight pressure on the bore of the sliding sleeve.

How does the baulk-ring device operate?

To engage the gear A the sliding sleeve is moved to the left by pressure applied through the gear shift lever. This move-

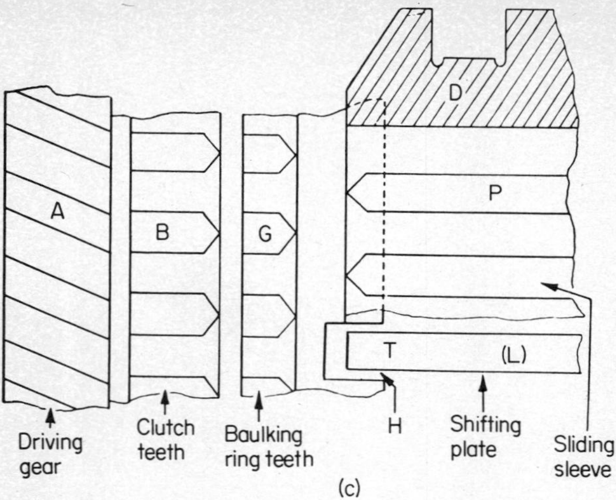

Fig. 43. (c) Synchro-hub synchronised position

ment will carry the shifting plates to the left to bring the baulk ring into light engagement with the mating cone C. At the same time, the gear A will be rotating at a speed which may be considerably different from that of the hub E and the baulk ring will therefore be rotated by friction to the extent permitted by the clearance H.

Reference to Fig. 43 will show exactly what has taken place. It will be noted that the baulk-ring teeth G have now moved into the path of the advancing sliding sleeve teeth P, which in turn endeavour to push the obstruction out of the way. In this manner, positive pressure is applied to the cone members, increasing the frictional effort between the cones. As further pressure is applied through the gear shift lever, the gear A is brought to the same speed as the sliding sleeve D. The frictional drag on the baulk ring is then zero and the chamfered teeth of the sleeve and the ring, acting on each other, force the baulk ring out of the way, thus allowing unrestricted movement into engagement with the clutch teeth

B. Since speeds are synchronised the movement is achieved in absolute silence.

What is the purpose of the grooves on the inner coned surface of the baulk ring?

The effect of the fine close pitch grooves is to cause a rapid breakdown of the oil film when the two members come together.

What limits the size of synchromesh devices?

The transfer of energy through the cone clutch is dependent on the size and weight of the objects to be accelerated or decelerated during a gear change. During each change the pinion to be selected, the layshaft, the input shaft and the clutch centre plate will increase or decrease in speed. The energy lost or gained via the cone clutches is absorbed in the form of heat.

What attempts have been made to increase the frictional loading?

A multi-coned synchromesh unit has been developed by Smiths of Cricklewood. Fig. 44 shows how two extra coned

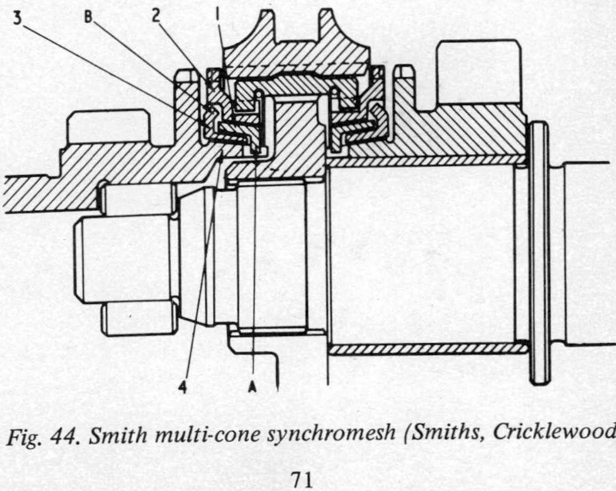

Fig. 44. Smith multi-cone synchromesh (Smiths, Cricklewood)

members (1) and (3) are positioned between the usual synchromesh cones (2) and (4). The extra members have a number of extensions A and B which engage in slots in members (1) and (3) to ensure that each part runs at the same speed as the shaft and gear. The frictional torque developed by the extra cones is added to that developed by the standard cones (2) and (4).

What advantages arise from this design?

Gear lever loads and synchronisation times can be reduced in normal sized (passenger car) gearboxes. It also opens up the possibility of designing units for commercial vehicles.

What other methods are in use?

Many other designs, embodying inertia-lock and baulk-ring principles, have been developed and used.

One of the most interesting is that of Dr Porsche, which can be described as a servo-synchromesh (Fig. 45). The sleeve has internal teeth which mate with the synchro-dogs mounted on the gear wheel. The teeth are cut back at an angle to form a coned surface at each end of the sleeve. On moving the sleeve towards the pinion to be synchronised, the internal teeth meet the coned surface of the split synchronising ring.

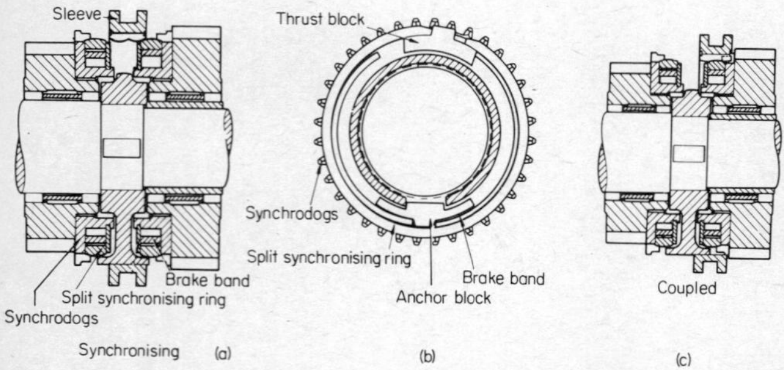

Fig. 45. Synchromesh (Porsche Cars)

This synchronising ring is prevented from rotating by the thrust block, butting against the brake band, positioned between the thrust block and the anchor block. The reaction of the synchronising force is taken by one of the brake bands which press against the inner surface of the synchronising ring, preventing it from collapsing until the gear dog is synchronised.

The self-servo action of this mechanism makes it very powerful in relation to the space it occupies. This feature makes the unit suitable for commercial vehicle gearboxes.

6

GEARBOX SELECTOR MECHANISMS

What methods are employed to move the gears or synchromesh units within the gearbox?

Two basic systems, floor change and column change, have become standardised. Floor change means that the gearbox has a lever extending from its upper face through the car floor. This is probably the most popular arrangement (see Fig. 50). Column change (Fig. 46) means that the gear selector lever is mounted on the steering column. This involves extra mechanical linkwork, giving rise to additional points for wear. Slackness in the joints will cause faulty gear selection.

How does the floor change operate?

The gear lever is usually mounted in a ball joint, so that the movement within the gearbox is opposite to that taken by the driver on selecting the gear. The lower end of the lever engages with a fork, mounted on a rail passing fore and aft through the gearbox casing (Fig. 47). Movement imparted to the gear lever is thus transferred to the rail and the fork assembly to move the synchro-unit or gear selector in the required direction.

How does the column change operate?

The Vauxhall control system is shown in Fig. 46. The gear shift lever and selector rod can be moved about the arc of a circle, so that a neutral is found in the mid position. To enable three speeds and a reverse to be selected, the lever and rod can be lifted bodily and then moved in the arc of a circle on this higher plane. Thus the selector levers and relay rods can be activated to select the required gear.

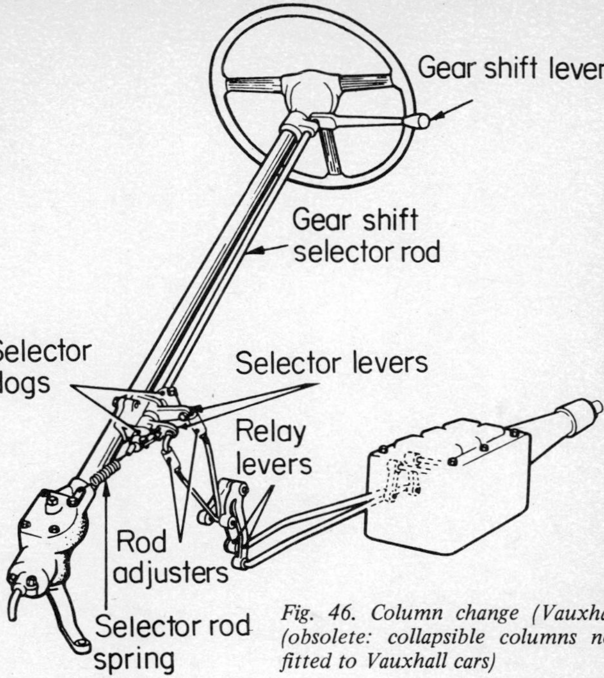

Fig. 46. Column change (Vauxhall) (obsolete: collapsible columns now fitted to Vauxhall cars)

What is the selector pattern?

To engage first gear, the gear selector lever is lifted from the lower neutral position to raise the selector rod against the lower spring tension, until the lower selector dogs engage with the selector lever. The lever is then moved anti-clockwise through the movement arc until the gear is engaged. The change to second gear is made by moving the gear lever to the neutral, where it is pulled down by the tension spring to the lower arc position, then moved clockwise to the engaged position. Top gear is engaged by moving the gear lever back through neutral without lifting the lever. To engage reverse,

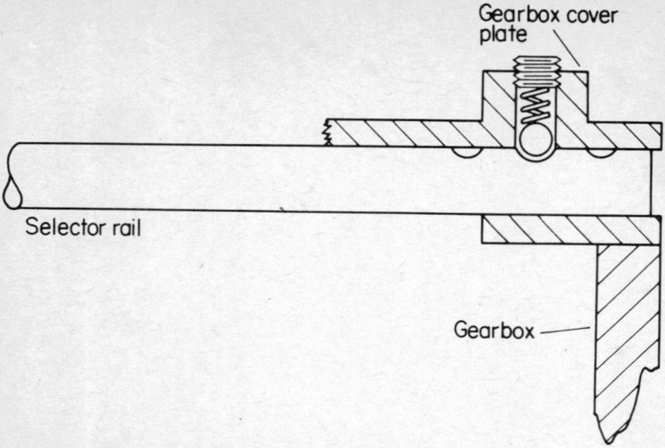

Fig. 47. Rail locking device (neutral position)

the gear lever is lifted in the neutral position, to engage the lower selector dogs, and then clockwise until the gear is effectively engaged.

What are the advantages and disadvantages of the column change?

Advantages: very little movement of the hand is required to reach the gear lever. More 'free' space or 'leg-room' available in the front compartment. Fitment of 'bench-type' front seat is possible.

Disadvantages: absence of positive gear select positions makes it difficult to master initially. Increased number of moving parts requiring lubrication. Greater cost to manufacture. Re-adjustment of the actuating mechanism may be necessary due to wear on the linkwork.

What holds the gear selector fork and rail in the engaged position within the gearbox?

The rail is usually held within the gearbox casing or cover plate with a spring-loaded plunger or ball bearing. The rail has suitable slots machined in it to accommodate the ball or plunger in the neutral and gear selected positions on either side (see Fig. 47).

What prevents two selector rods being moved at the same time?

In a four-speed gearbox, using a three-rail system, a ball lock arrangement, as shown in Fig. 48, may be employed. Diagram (b) shows the centre rail moved away from neutral, with the outer rails locked by the ball bearings. Diagram (c) shows one outer rail moved away from neutral, causing one ball bearing to force the sliding pin through the centre rail to lock the remaining rails to the gearbox casing.

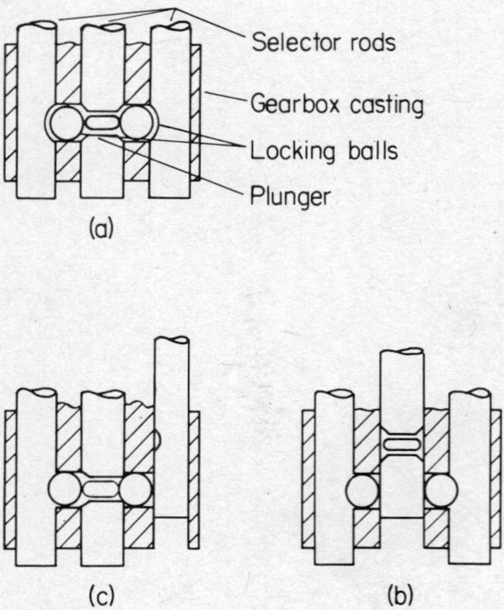

(a) Neutral. (b) Centre rod moved. (c) Outer rod moved

Fig. 48. Three-rail selector interlock

What method is employed in a three-speed gearbox?

One ball bearing can be used between the rails as shown in Fig. 49. The diagram is self-explanatory.

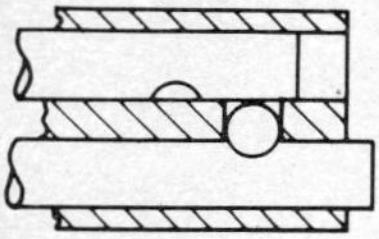

Fig. 49. Two-rail interlock

Is there an alternative interlocking device for a three-rail selector system?

A simple caliper, as shown in Fig. 50, will lock two selector rails. The gap in the caliper follows the gear selector lever

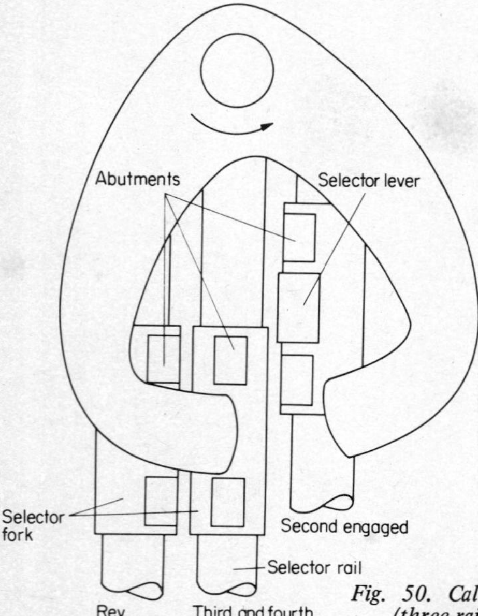

Fig. 50. Caliper interlock (three-rail selector)

across the neutral position, automatically locking the other rails at any given point. When a gear is selected, so that the gear lever is moved out of the neutral position, the side of the fork retains the caliper in the lock position.

Is there an alternative interlocking device for a two-rail selector system?

A simple and ingenious interlock is used in the Vauxhall gearbox (Fig. 51). The two rails A and B carry the selector forks and are positioned on either side of the gear wheel

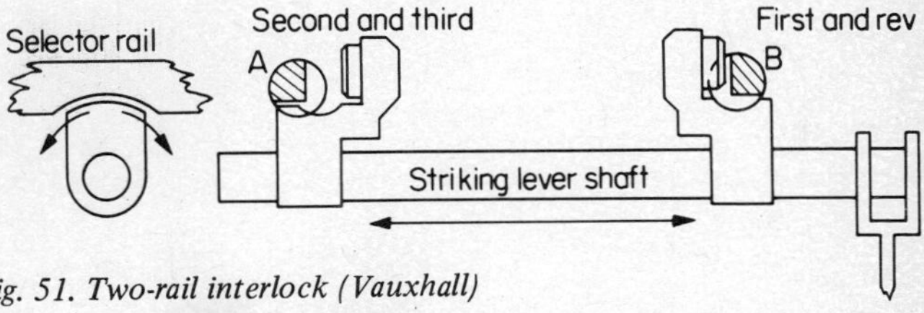

Fig. 51. Two-rail interlock (Vauxhall)

cluster, at about half the depth of the gearbox. A striking lever shaft passes transversely through the box, below the rails, and is free to move axially within the limits of the buttons on the striking levers. It is also able to rock fore and aft to move the rails A or B to select the required gear ratio. In the position shown the rail B is selected to engage first or reverse gear as required by the driver, while the striking button of the other lever is clear of the rail. It should be noted that the hub of the left-hand striking lever is engaged with a concave recess in rail A to effect an interlock, which does not prevent the rocking action of the striking lever shaft.

How does the single-rail selector mechanism, as fitted to the Ford Escort, provide an interlock?

The selector rail (Fig. 52) carries the first and second, and the third and top selector forks. These are free to slide on the rail. The reverse selector fork pivots on its own fulcrum pin in the gearbox housing and engages with the reverse idler gear.

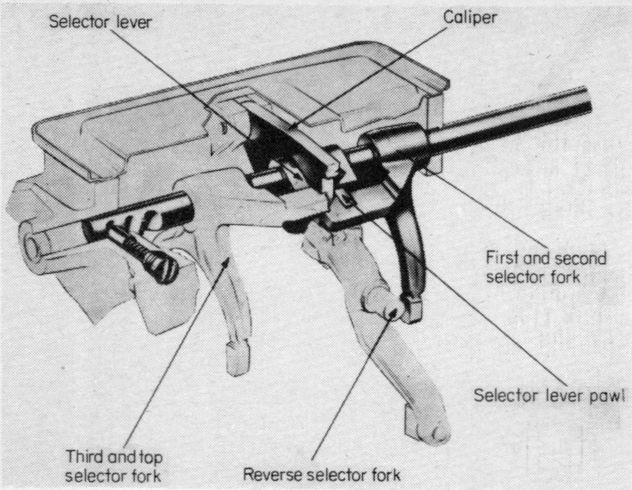

Fig. 52. Escort single-rail interlock (Ford)

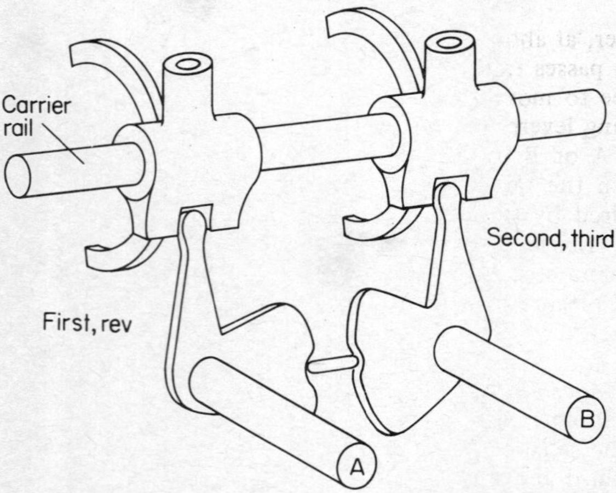

Fig. 53. Two-lever interlock for column change

The selector lever is fixed to the rail and rocks about the arc of a circle as the gear lever is moved from side to side. The selector lever pawl picks up the individual forks as required. In the figure, first gear is engaged. The selector lever actuates a caliper (pivoted to the gearbox casing), positioning it to engage with the slots in the selector forks, not in engagement, thus effecting an interlock.

What method of interlocking is used in gearboxes, having selector levers in conjunction with a column change?

Fig. 53 shows a single rail carrying two selector forks, which are free to move along the rail within the limits of the required travel. Ball locks are provided for each engaged position. Two rocking levers A and B, connected to the gear change lever, provide the motion to each fork. The levers are constructed with a cam section to obtain the interlock effect. A single plunger mounted in the gearbox casing between the cams is of sufficient length to allow only one cam at a time to be operational.

7

OVERDRIVES

What is the case for fitment of an overdrive gear?

Fig. 9 shows the relationship between road speed and power. The maximum road speed is beyond the point of maximum power, to make available a reasonable amount of power for acceleration throughout the speed range. The power required to drive the car at maximum speed is less than the maximum power available from the engine. It should be noted that this power is also available on the rising side of the power curve.

What advantages may be claimed for the use of an overdrive?

From Fig. 9 it can be seen that the power required for maximum road speed in top gear can be obtained in overdrive on the rising side of the power curve with the engine turning at a considerably reduced speed. This contributes to reduced wear on the engine and a saving in fuel used.

What type of gearing is used to obtain the overdrive ratio?

Two designs are possible, (a) a normal helical gear set, with its own clutch and synchromesh unit, or (b) an epicyclic gear with a brake band or cone clutch. The majority of overdrives incorporate the epicyclic gear, which requires no gear lever or synchromesh unit.

Do all overdrives fit the same type of epicyclic gear?

No. The Borg-Warner overdrive uses a simple epicyclic gear (Fig. 63), while the Laycock-de-Normanville unit (Fig. 56) uses compound planet wheels.

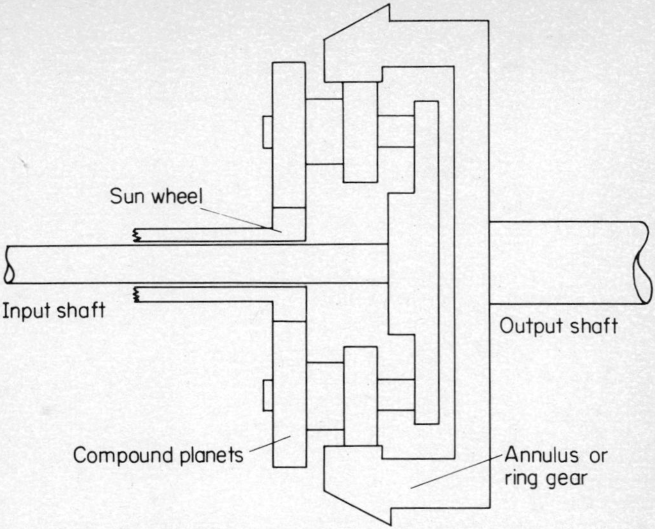

Fig. 54. Compound planet gears (Laycock overdrive)

Why are compound planets used in the Laycock overdrive?

To obtain a large sun wheel, so that the drive shaft may pass through the boss; the gear ratio is arranged in two steps, by means of the compound planets (Fig. 54).

What is the power transmission line in direct drive and in overdrive?

Fig. 55(a) shows the unit in direct drive; the double cone clutch and the roller one-way clutch have been added. A simple epicyclic gear train is used in this illustration for simplicity. In direct drive power is applied to the input shaft from the gearbox to the planet carrier through the uni-directional clutch to the outer race, which is attached to the output shaft. In this condition the cone clutch is held by the spring pressure on to the ring gear and the sun wheel is held as a unit by virtue of the mating splines.

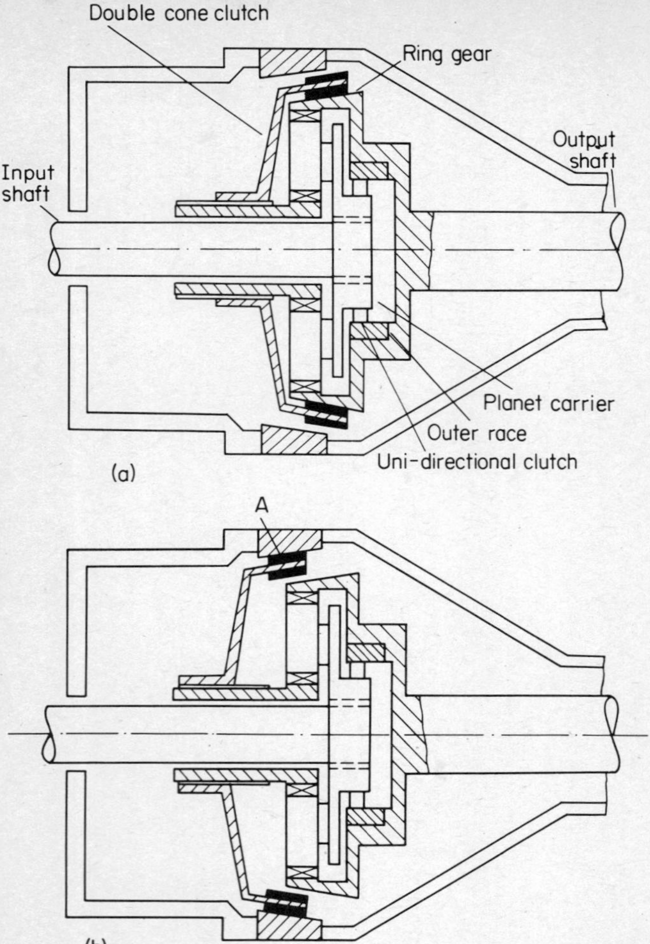

(a) Direct drive. (b) Overdrive

Fig. 55. Direct drive and overdrive

In overdrive (Fig. 55 (b)) the friction cone is moved by hydraulic pressure against the spring load to engage with the overdrive casing at A. The sun wheel is now held stationary. Power enters the unit from the gearbox to the planet carrier, causing the pinions to revolve about the stationary sun. This action transmits the drive to the output shaft via the ring gear. (Fig. 56 shows a fuller representation of the assembly.)

What other duties does the inner cone perform when the unit is in direct drive?

To give engine braking on a change down from overdrive to direct drive, and to drive the vehicle in reverse, when the roller clutch is inoperative.

What maintains the drive during a change from one clutch surface to the other?

The drive is maintained through the roller one-way clutch, unless the propeller shaft is overrunning or 'free-wheeling'.

Can the unit 'free-wheel' in the overdrive condition?

No. Free-wheeling is only possible in direct drive.

How is the spring load applied to the clutch member when the unit is in direct drive?

The coil springs do not revolve with the friction cone. The cone and the sun wheel are supported by a ball race which is attached to the inner thrust member of the spring cluster. The opposite end of the coil springs are located in the overdrive casing.

How is the clutch member moved to the overdrive position?

By hydraulic pressure moving the operating pistons within the cylinder bore in the main casing, so that they move the cone clutch against the spring load (Fig. 56). The operating pressure is about 275 N/cm^2. The outer friction surface now makes contact with the casing of the unit, so that the clutch member and the sun wheel are now stationary. This action allows the sun wheel to be used as a reaction member, bringing the overdrive gear into operation.

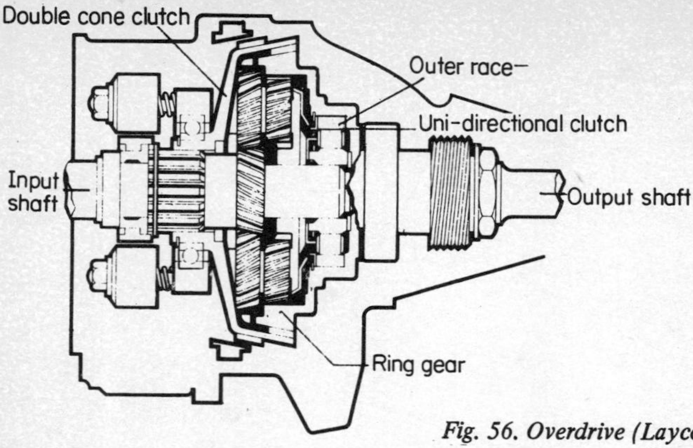

Fig. 56. Overdrive (Laycock)

How is the hydraulic pressure provided?

A small plunger pump (Fig. 57) is driven from the input shaft by means of an eccentric keyed to it. Pressure is available at all times when the vehicle is in motion.

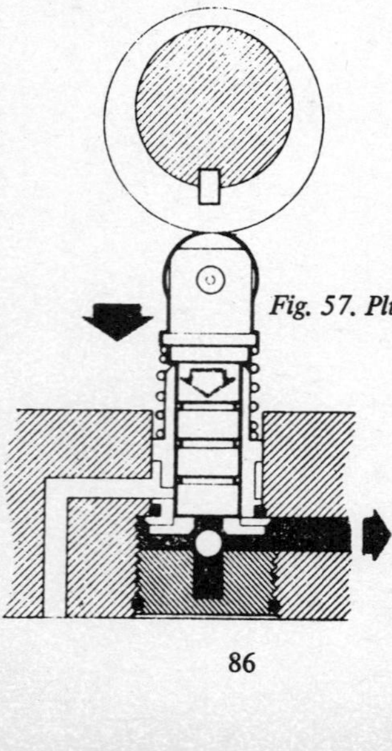

Fig. 57. Plunger type pump (Ford)

How does the pump operate?

Motion is provided by the eccentric forcing the plunger down into the pump body; the return stroke is accomplished by means of the return spring. On the up stroke, with the 'foot' valve closed, a partial vacuum set up by the piston causes oil (initially supplied from the main gearbox) to enter through the inlet port. On the down stroke, as soon as the inlet port is closed a pressure is created, opening the foot valve to allow the oil to pass into the hydraulic circuit.

What constitutes the complete hydraulic circuit?

The complete system is shown in Fig. 58. In the overdrive condition, the plunger pump supplies oil under pressure to the regulating valve, the solenoid valve and the operating cylinders

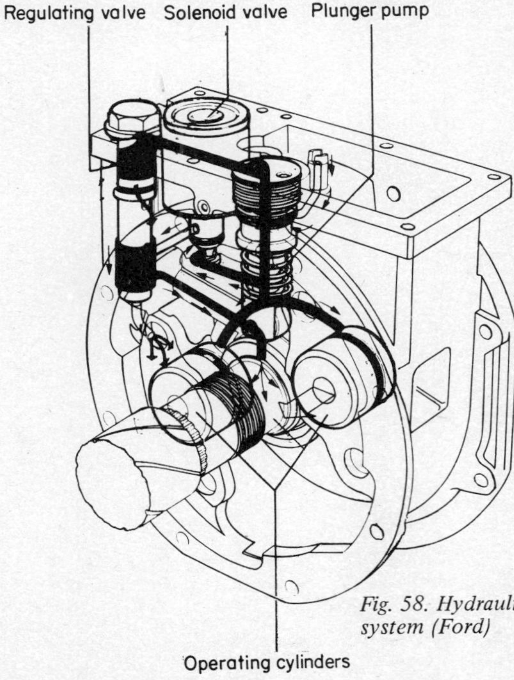

Fig. 58. Hydraulic control system (Ford)

simultaneously. The action is triggered by the solenoid valve, as and when it is actuated by the electrical control gear.

How is the electric control arranged?

A simplified wiring diagram is shown in Fig. 59. The column switch is closed to activate the solenoid valve D (see Fig. 60 for operation of the valve). The circuit is supplied from the SW terminal B on the ignition coil through the relay unit C between the terminals W_1 and W_2. To make certain that the overdrive can only operate in, say, third and top gear, an isolator switch F is closed by the gearbox selector mechanism.

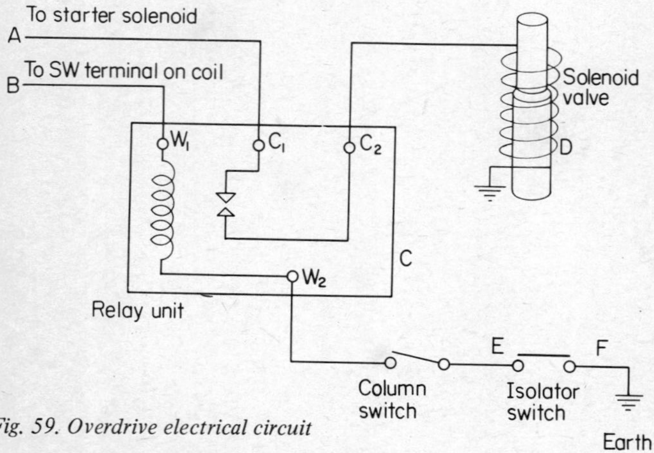

Fig. 59. Overdrive electrical circuit

The energised coil of the relay closes the points across the C_1 and C_2 terminals to complete the operational circuit from the starter solenoid to the overdrive solenoid D.

What does the solenoid valve do?

When overdrive is selected, the coil (Fig. 60) is energised and the armature moves upwards to press the ball into the valve body to restrict the fluid flow. Pressure then builds up in the circuit.

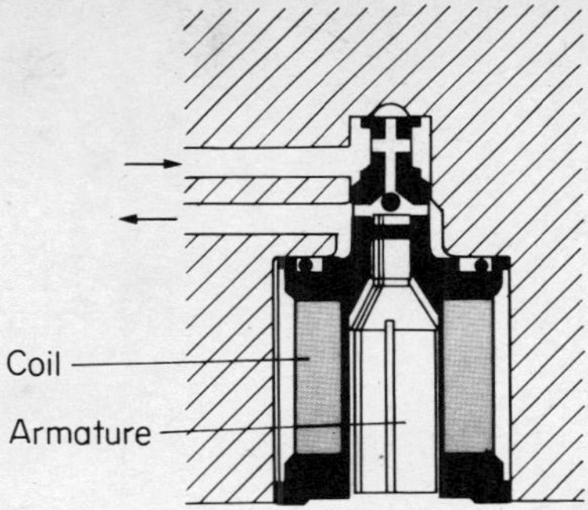

Fig. 60. Solenoid valve (Ford)

How is the pressure in the hydraulic circuit controlled?

It is controlled by the relief valve, which in effect is a double valve (see Fig. 61). When direct drive is engaged the oil under pressure from the pump is used for lubrication only and returns to the sump. The oil enters through the port X and

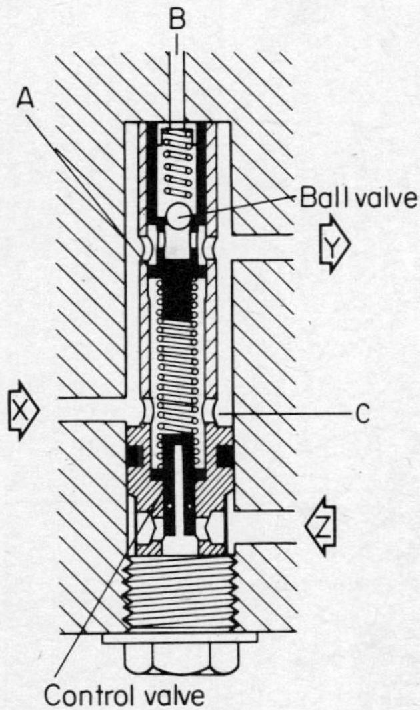

Fig. 61. Relief valve (Ford)

rises in the valve body through the port A and out via the lubrication port Y. This low pressure is monitored by the ball valve. Any rise in pressure caused by obstruction in the lubrication system, or increase in engine rev/min, lifts the ball off its seating to allow 'dumping' through the port B.

When the overdrive is selected and the solenoid valve is closed, oil under pressure is also fed into the valve through the port Z, passing through a fine gauze filter, to act on the control valve. The control valve is set to a predetermined high pressure. If the pressure rises above this level, the valve lifts to allow 'dumping' into the main body and on into the lubrication circuit through port C.

How is the clutch cone moved by the hydraulic pressure?

Two pistons working in cylinders machined in the overdrive casing push on two tie-bars attached to the studs screwed into the thrust member of the clutch (see Fig. 62).

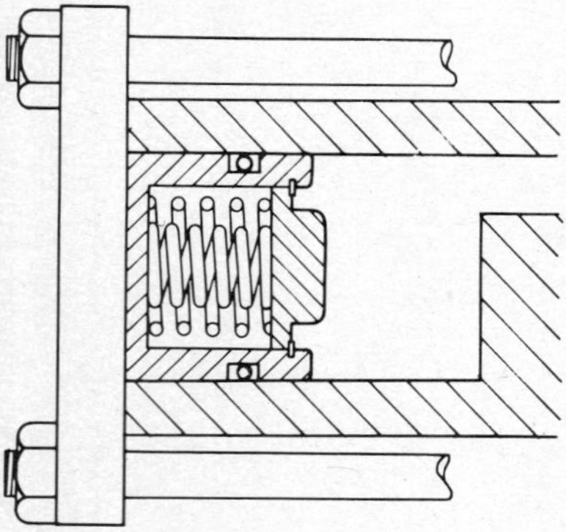

Fig. 62. Thrust cylinder

What is the purpose of the springs and plunger within the pistons?

When the pistons have moved to the extent of their travel and the pressure continues to rise, the damping plungers compress the internal springs; this increase in pressure should signal the high pressure valve to open. Slight variations in pressure will now be absorbed by the movement of the damping plungers and NOT by the piston body.

What checks can be made on the Laycock overdrive?

Very little can be done to the unit without dismantling, but if the overdrive fails to operate, first check the electrical circuit with a voltmeter or test lamp. The solenoid lead should be 'live' when the overdrive switch is closed and the appropriate gear selected.

See that there is sufficient oil in the gearbox, and top up if required. Should the level in the gearbox be correct, the only remaining test will be for pressure in the oil system. Pressure should be checked in overdrive with the engine running and the gear selected. For static testing this will require the car to be on safe axle stands. At 48 km/h the pressure should not be less than 275 N/cm^2. Subsequent action will depend on the results of the above tests.

Is there an overdrive without hydraulic control?

The Borg-Warner overdrive is controlled by a solenoid operated pawl, or it can be locked out by a separate mechanical linkage.

What type of gearing is used in this overdrive?

A simple epicyclic gear train is used. The sun wheel, planet wheels and the annulus or ring gear can be readily seen in Fig. 63.

How does the gearset operate in direct and overdrive positions?

In direct drive the sun wheel is unlocked and the epicyclic gear is rendered inoperative. Power is transmitted through the roller clutch to the output shaft. This clutch also allows 'free-wheeling' at speeds below 56 km/h. In overdrive the sun wheel is locked, so that the power is transmitted from the gearbox mainshaft to the planet wheels, via the carrier. The planet wheels rotate on the stationary sun wheel to turn the annulus and the output shaft in the higher ratio.

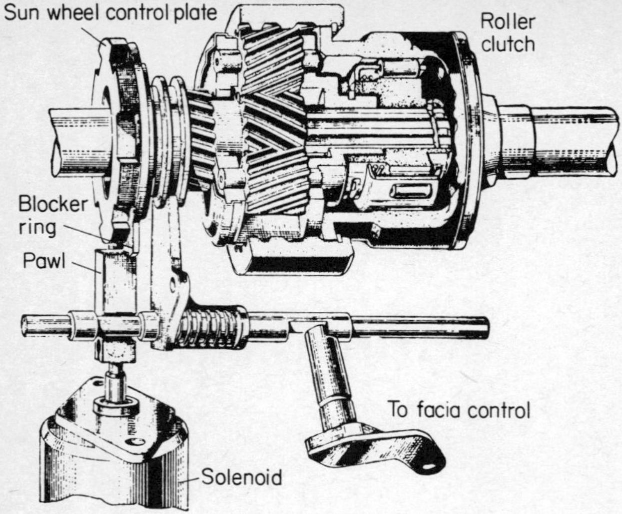

Fig. 63. Overdrive (Borg-Warner)

Can the overdrive be 'locked out' to provide engine braking?

Yes. A facia control is provided to move the sun wheel along the shaft, to engage it with the planet carrier. The drive is now taken through the locked up epicyclic gear, instead of passing through the free-wheel.

This lock is applied automatically when reverse gear is selected owing to the free-wheel being unable to transmit the drive in the reverse direction (see Fig. 63).

What is the normal sequence of selecting the overdrive ratio?

At speeds above approximately 56 km/h the overdrive becomes available automatically, by means of an electrical governor switch driven from the output shaft and therefore sensitive to road speed (Fig. 63). It energises a solenoid (through a relay switch), which engages a pawl with the sun wheel. The actual engagement of the pawl will not take place until the accelerator pedal is released momentarily.

What effect has the accelerator on the engage or disengage sequence?

To engage the overdrive the accelerator pedal is released momentarily to allow the engine revolutions to drop to an appropriate level to ensure shock-free engagement. The overdrive can be disengaged by means of the 'kickdown' switch. Full depression of the pedal opens one set of contacts to break the solenoid circuit; while at the same time a second set of contacts (within the solenoid) cause the ignition circuit to be momentarily interrupted to relieve the pawl of torque loading, to ensure immediate withdrawal (see Fig. 64).

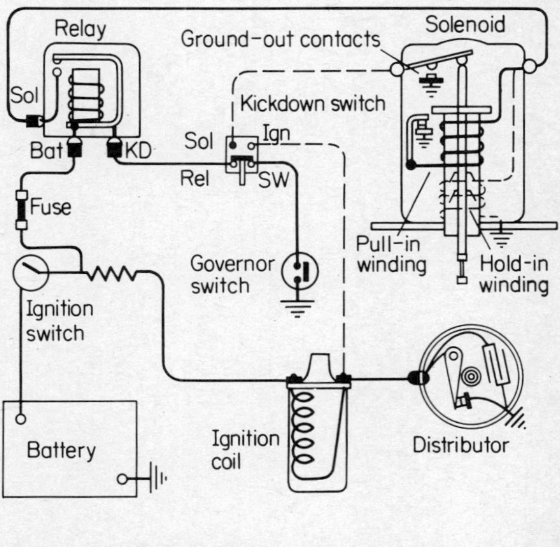

Fig. 64. Overdrive wiring diagram (Borg-Warner)

What controls the automatic application of the overdrive?

The automatic engage and disengage is controlled by an electrical circuit, operating suitable relays and solenoid. The electrical governor switch (Figs. 63 and 64) completes the circuit to energise the solenoid controlling the locking pawl, which holds the sun wheel in overdrive. The accelerator pedal must be momentarily released, otherwise the pawl will not engage. When the gear is selected, the pull-in winding of the solenoid is switched off and the hold-in winding is energised. Normal disengagement occurs when the road speed falls below 50 km/h. The solenoid will again interrupt the ignition system to allow the release of the pawl (see Fig. 64). The interruption of the ignition circuit is not noticeable when driving the car.

Have any overdrives been built into the gearbox?

Yes, a number of manufacturers have done this. Fig. 65 shows the Austin Maxi arrangement. The gear trains are marked 1st, 2nd, 3rd, 4th, 5th and R. The reverse control lever is

Fig. 65. Built-in overdrive (British Leyland)

attached to the side of the gearbox and is therefore removed from the illustration. The right-hand wheel on the layshaft is the overdrive gear meshing with the final drive pinion at the end of the mainshaft. The overdrive gear is mounted on needle rollers, running on an extension to the layshaft, so that in the normal gearbox ratios it is running free. To engage the overdrive for power transmission the gear is locked to the layshaft by a baulk-ring synchro-unit, whose hub is splined to the shaft and retained in position by a nut and tab washer.

8

FOUR-WHEEL DRIVES

How is the drive-line split for four-wheel drive?

Two arrangements are possible, (a) a transfer box making four- or two-wheel drive available, and (b) a four-wheel drive, as in the Jensen FF system in which two-wheel driving is not possible.

How does the transfer box operate?

Fig. 66 shows a transfer box, as used in the Land Rover, bolted to the rear of the normal gearbox. It provides a high and low ratio, in addition to the four speeds of the normal gearbox, and means of engaging and disengaging the front axle to give two- or four-wheel drive at the will of the driver.

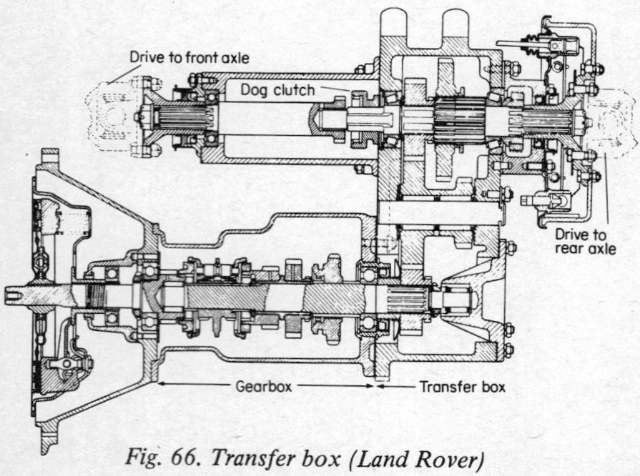

Fig. 66. Transfer box (Land Rover)

How are the additional high and low gears achieved?

In Fig. 66 the idler wheel in the transfer box (connecting the mainshaft with the propeller shaft drive) has two gear wheels; the smaller gear can be meshed with the sliding wheel on the transfer shaft to give a reduction, or low gear ratio. When the sliding mesh wheel moves to the left it forms a dog clutch engagement with the constant mesh wheel in the transfer box, providing a high or overdrive ratio. These additional gears are independent of the main gearbox.

How is the front-wheel drive controlled?

In the transfer box the front axle drive shaft is mounted on a spigot bearing, while a dog clutch is splined to the rear axle drive shaft. In the position shown, the drive to the front wheels is disconnected. Movement of the dog clutch to the left will engage the mating teeth to give four-wheel drive.

What are the requirements for four wheels driving at all times?

In order to obtain a drive to all the wheels all the time, a centre differential is required, so that the torque can be split in proportion to the load on the front and rear axles. The epicyclic gear (top right) in Fig. 67 forms the centre differential of the Jensen FF drive. The input to the planet carrier is split by the gear ratio sun-wheel-to-annulus. This torque split ratio can only be correct for a given set of operating conditions, so that it is necessary to control the action of the centre differential to avoid wheel spin occurring.

The centre differential must also allow a degree of freedom between front and rear axles during cornering under drive torque conditions existing at each road wheel. A permissible speed difference is built into the control system.

How does the Jensen 4WD unit operate?

The 4WD unit replaces the tailshaft housing of the automatic gearbox, the output of the gearbox becoming the input of the unit as shown in Fig. 67. This shaft drives the planet carrier of the epicyclic differential, the ring gear or annulus being an integral part of the rear output, or drive to the rear

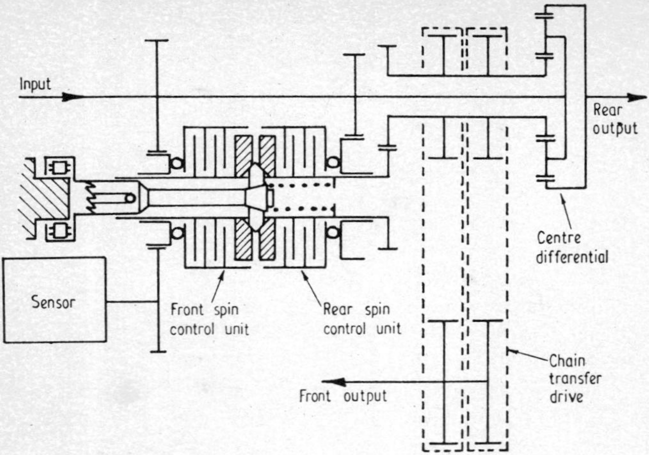

Fig. 67. Jensen FF four-wheel drive (Institution of Mechanical Engineers from Drive Lane Engineering)

axle. The sun gear drives the front output shaft by means of two inverted tooth chains. The front and rear axle ratios are equal, so that the input and output shafts run at the same speed when the vehicle is running in line ahead and no differential action occurs. The torque output is split unequally by the centre differential, which transmits 63% to the rear axle and 37% to the front axle.

The differential action is limited by one-way clutches mounted on the mainshaft of the control unit. This mainshaft is driven by gearing from the sun gear shaft. The clutch drums are driven by a pair of gears on the input shaft, the left-hand one being driven faster and the right-hand one slower than the mainshaft.

What is the action of the control clutches?

Each clutch is made up of alternate steel and phosphor bronze plates. The steel plates are splined to the mainshaft while the bronze plates are tongued into the clutch drums, each pack of plates being located axially by abutment rings, whose conical bores bear against chamfered ends of radial plungers. An inner sleeve prevents the plungers from pressing inwards by the thrust of the abutment rings. Three compression springs between the abutment rings make certain that the clutch plates are held in light contact. Between the clutch drums and the control gears are three steel balls held in helical ramps between the opposing faces.

Under normal running conditions the clutch drums run at the same speed as the control gears, without exerting any significant force on the slipping clutch packs. This is achieved by the left- or right-handing of the helical ramps. Neither clutch can transmit any torque until the front or rear wheels tend to spin or lock. If the front wheels tend to spin the speed increase of the mainshaft is greater than that of the left-hand clutch drum. When the speeds coincide the reversal of the clutch drag rolls the steel balls up the ramps. A rapid servo action takes place, so that the clutch pack, the mainshaft and the control unit lock together, preventing further differential action of the epicyclic train, transferring the excess front drive torque to the rear output shaft.

When the clutch outer member slows down to the speed of the mainshaft, the grip of the clutch plates prevents the rear wheels locking, and excess rear braking effort is transferred to the front wheels. The right-hand control unit prevents the rear wheels slipping and the front wheels locking in a similar manner.

What happens if all the wheels lock or tend to spin?

No single wheel can lock or spin independently, as this would cause a change in propeller shaft speed and the control units would be put into operation. The system cannot prevent all wheels locking or spinning. In the event of all wheels tending to lock a skid sensing unit, capable of distinguishing between normal deceleration due to braking and rapid deceleration of a locking wheel, is arranged to release the brakes sufficiently to maintain wheel rotation.

What is the effect on the system when the vehicle is reversing?

All the shafts in the 4WD unit rotate backwards and the control clutches tend to lock. This action is prevented by a sprag type one-way clutch, to the left of the mainshaft, with its inner track fixed to the casing. The outer track, unable to rotate backwards, forces the inner sleeve by means of face ramps to move to the right. This action causes the radial plungers to move inwards, allowing the abutment ring to close together. Under these conditions no servo action is possible at the ball ramps to lock up the clutches.

When forward motion is resumed, the central spring causes the inner sleeve to move to the left and the abutment rings are restored to their original position by the radial plungers moving away from the shaft centre. The control unit is again set for normal operation.

9

AUTOMATIC TRANSMISSIONS

What is meant by the term 'automatic transmission'?

This is usually a gearbox that will select the appropriate gear ratio automatically once the driver has selected a forward or reverse drive at the control lever.

What method is used to indicate the gear pattern?

A control lever moving across a quadrant is often used. The quadrant of the basic Borg-Warner gearbox is marked P.R.N.D.L. Additional drive conditions may be indicated by the omission of the letter L and substituting numbers 1 and 2 to denote alternative drive patterns.

What is the meaning of the basic lettering?

P is for Park. This selects a pawl in the gearbox to engage with the output shaft to provide a mechanical lock against rotation. This condition must only be selected when the vehicle is at rest. This is also one of the select positions in which the engine may be started.

R is for Reverse, selecting a single-speed reverse with engine braking. (No up- or down-shifts can take place.)

N is for Neutral, when the drive from the engine is disconnected. This is the second condition in which the engine can be started.

D is for Drive, with normal forward motion, changing up and down, depending on the assessment of the gearbox in terms of speed and load on the engine, together with the line pressure in the hydraulic control system.

L signifies Lock-up. In this condition only first and second gear are available. If the vehicle is in third gear and L is selected, the box will immediately move to second; and then providing the speed drops sufficiently, first gear will be engaged and the box will remain in this condition until the selector lever is moved out of L. Thus, down shifts only are available, depending on road speed.

In the alternative pattern, when L is omitted from the quadrant, 2 provides for drive in first and second gear, with up and down changes depending on the prevailing conditions. If 2 is selected when the car is moving in third, an immediate downshift to second will take place. 1 provides for lock in first, the transmission changing down to this ratio as and when the road speed is low enough. The transmission will remain in first gear until the selector is moved out of 1.

Is this a standard nomenclature for all makes of transmissions?

No. Individual manufacturers have their own markings, but in general the letters have the same meaning throughout. Additional numbers may appear, as in the Automotive Products transmission R.N.1.2.3.4.D. (see page 109).

Do all automatic transmissions have the same gearset?

No, but the majority of the makes use one of three gearsets, namely, the Simpson gearset, the Ravigneau system (as in Borg-Warner), and the Reverted epicyclic, which is the main feature of the Automotive Products transmission as fitted to B.L.M.C. cars.

What is the Simpson gearset?

This gearset is shown in Fig. 68. It can be identified by the common sun wheel between the gear trains. All the planet wheels are the same size, and the ring gears have the same number of teeth.

How many gears does this system provide?

Three forward and one reverse. The intermediate gear ratios will depend on the number of teeth on the sun wheel, the planet wheels and the ring gear. If the number of teeth on the wheels are 30, 18 and 66 respectively, the ratios will be 2.45:1 in low, 1.45:1 in intermediate and 1:1 in top. Reverse provides a 2.2:1 reduction.

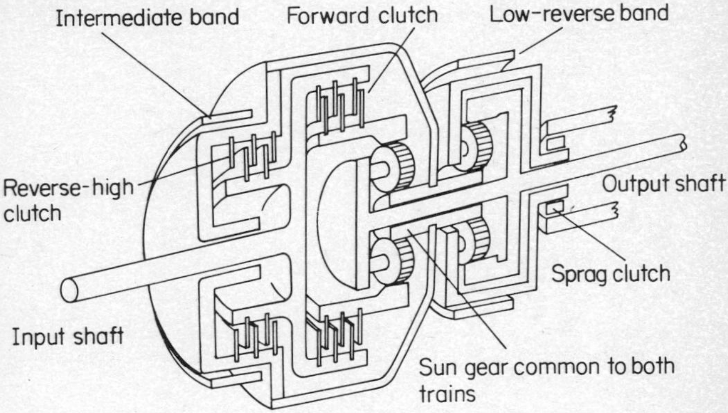

Fig. 68. Simpson gearset

How are the ratios obtained?

The selection of ratios depends on the action of clutches and brake bands operating to a set pattern. Low gear in the normal drive sequence is obtained by applying the forward clutch (Fig. 68). Power is applied to the ring gear of the first set, rotating the planet wheels clockwise to drive the common sun wheel anti-clockwise. The planet carrier of the second set is held against rotation by the one-way clutch. Thus the planet wheels rotate to drive the output shaft in compound reduction. To provide engine braking in first gear, the low-reverse band is applied to cut out the free-wheel effect of the one-way clutch.

Intermediate or second gear is obtained by applying the intermediate band to hold the sun wheel against rotation. Power is applied through the forward clutch to the first ring gear to drive the planet wheels around the stationary sun, to rotate the output shaft by simple reduction.

Top gear is obtained when the forward and reverse-high clutches are engaged, to lock the common sun wheel to the first ring gear. The gearset revolves as one unit at the same speed as the input shaft.

To select reverse, the forward clutch is released and the reverse-high clutch and the low-reverse band are applied. The reverse-high clutch transmits power to the sun gear via the input shell, to rotate the planet wheels of the second gear train, turning the ring gear in the opposite direction to the input shaft. The planet carrier of the second gear train is held against rotation by the brake band.

How are the gear ratios of the Simpson gearset calculated?

A number of methods to determine the ratio of epicyclic gearing have been developed. In this discussion the formula

$$e = \frac{N_l - N_a}{N_f - N_a}$$

will be used, where

e = Gear ratio
N_f = Speed of first shaft (rev/min)
N_l = Speed of last shaft (rev/min),

given that the number of teeth on the sun wheel is 30, while all the planets have 18 and each ring gear 66 teeth, as in the Chrysler Torqueflite transmission. The ratios can be calculated as follows.

Low gear. With the planet carrier of the rear epicyclic held against rotation, this train is in simple reduction of $\frac{66}{30} = 2.2:1$.

A shaft speed must be assumed to proceed with the calculation. Taking an output shaft speed of 1000 rev/min, the sun wheel will revolve at 2200 rev/min anti-clockwise. Thus the gear ratio should be written $e = -\frac{66}{30}$ or -2.2.

The speed of the first shaft can now be found:

$$-2.2 = \frac{N_1 - N_a}{N_f - N_a}$$

$$-2.2 = \frac{-2200 - 1000}{N_f \quad - 1000}$$

$$= -2.2\,N_f + 2200 = -3200$$

$$\text{or } 5400 = 2.2\,N_f$$

$$N_f = 2450$$

The gearbox ratio in first gear is $\frac{2450}{1000}$ or 2.45:1.

Intermediate or second gear. In this gear the common sun wheel is held against rotation. Power applied to the first ring gear drives the planets around the stationary sun to rotate the output shaft by simple reduction.

$$e \equiv -\frac{66}{18} \times \frac{18}{30} = -2.2$$

Assuming in this instance that the input speed is 1000 rev/min the speed of the arm can be found, the arm being in effect the planet carrier.

$$e = \frac{N_1 - N_a}{N_f - N_a}$$

$$-2.2\,(1000 - N_a) = 0 - N_a$$

$$-2200 + 2.2\,N_a = 0 - N_a$$

$$3.2\,N_a = 2200$$

$$N_a = \frac{2200}{3.2} = 690 \text{ rev/min}$$

$$\text{Gear ratio} = \frac{1000}{690} = 1.45:1$$

Top gear. This is direct drive 1:1.

Reverse gear. The planet carrier of the second or rear gear train is now held against rotation. The drive enters via the common sun wheel to provide a simple gear train

$$\frac{18}{30} \times \frac{66}{18} = 2.2{:}1$$

What type of gearset is used in the Borg-Warner transmission?

The Model 8 and the Model 35 transmission use the same type of gearing. This gearset (Fig. 69) has two sun wheels, two sets of planet wheels, a planet carrier and a ring gear, providing three forward speeds and a reverse gear.

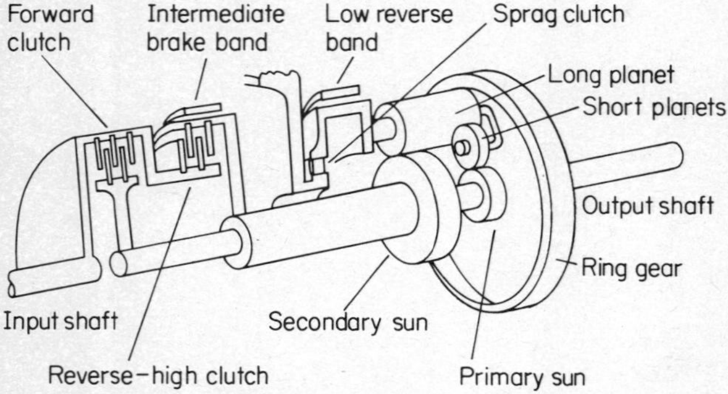

Fig. 69. Borg-Warner gearset

What is the reason for the two distinct models?

The Model 8 Borg-Warner transmission is virtually a scaled up and more robust version of the Model 35 unit, for use on cars with 3 – 4 litre engines. The Model 35 may be fitted to comparatively small engines from 1 litre upwards.

How are the gear ratios obtained?

In first gear the planet carrier is held stationary, either by the brake band (in lock-up) or by the one-way clutch (in drive selected). The power is applied to the primary sun, via the forward clutch, so that the short planets (16 teeth) rotate anti-clockwise. The planet carrier is held, therefore the long planets (17 teeth) rotate clockwise to drive the ring gear (67 teeth) in the same direction as the input shaft. This gear provides a reduction ratio of 2.39:1.

Intermediate or second gear is brought into action by the application of the intermediate brake band, in addition to the forward clutch. The secondary or reverse sun wheel is held stationary, while the planet carrier is released, so that it can revolve freely. Power is applied to the primary sun, as in first gear, the two sets of planet wheels now rotate and the long planets revolve on the stationary sun wheel. In so doing they turn the annulus in a reduction ratio of 1.45:1.

Third or top gear is obtained by releasing the intermediate brake band and applying the reverse-high clutch in addition to the forward clutch already applied. The action of the two clutches locks both sun wheels to the input shaft. The gears now rotate as one unit, providing a direct drive.

For reverse, the forward clutch is released and the reverse-high clutch is engaged, together with the low-reverse band. This is necessary because the one-way clutch will not hold the planet carrier in this ratio. Power is applied to the secondary or reverse sun wheel, which rotates the long planets on their axis, to turn the annulus (or ring gear) in the opposite direction to the input shaft. The ratio in this gear is 2.09:1.

How are the gear ratios of the Borg-Warner Model 35 transmission calculated?

First gear. In this gear there is no epicyclic action, as the planet carrier is held against rotation and the gearing resolves itself to a simple gear train.

$$\frac{16}{28} \times \frac{17}{16} \times \frac{67}{17} = \frac{67}{28} = 2.39:1.$$

Intermediate or second gear. Here the reverse sun is held against rotation and the planet carrier is released. Thus the action of the gear is epicyclic. The gear wheels in this system are not of the same module, so that another method is used to find the gear ratio.

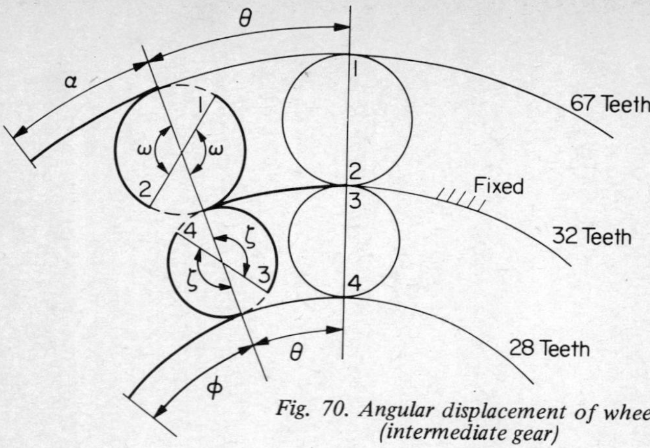

Fig. 70. Angular displacement of wheels (intermediate gear)

From Fig. 70, where the thick lines show the angular displacement of each gear wheel:

$$32\theta = 17\omega = 67\alpha = 16\zeta = 28\sigma$$

Using N = rev/min and number of teeth on each wheel for identification, the gear ratio is

$$\frac{N_{28}}{N_{67}} = \frac{\sigma + \theta}{\alpha + \theta}$$

From the above $\sigma = \dfrac{32}{28}\,\theta$ and $\alpha = \dfrac{32}{67}\,\theta$ so that

$$\frac{N_{28}}{N_{67}} = \frac{(32/28)\,\theta + \theta}{(32/67)\,\theta + \theta} = \frac{\theta\,(32/28) + 1}{\theta\,(32/67) + 1}$$

$$= \frac{32 + 28}{28} \times \frac{67}{32 + 67}$$

$$= \frac{60}{68} \times \frac{67}{99} = 1.45{:}1$$

Top gear. Top gear is a direct drive of 1:1.

Reverse. With power applied to the reverse sun wheel, with the planet carrier held against rotation, no epicyclic effect takes place, giving a simple gear train of $\frac{17}{32} \times \frac{67}{17} = \frac{67}{32} = 2.09:1$.

What is a reverted gear train?

This is a gear train in which the first and the last gears are on the same axis. A simple example can be formed by two shafts connected by a differential gear contained within a housing. (See Fig. 84.)

How is this type of gearing employed in the Automotive Products transmission?

Reference to Fig. 71 will **show** what appear to be two differential gearsets, one inside the other.

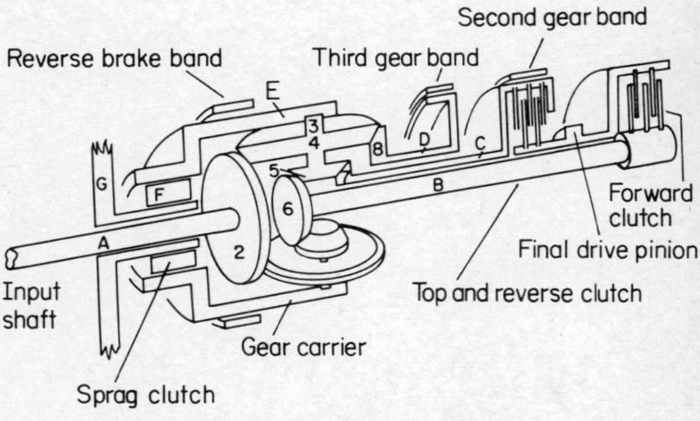

Fig. 71. Automotive Products transmission

How are the gears brought into operation?

Shaft A is the input shaft driven by a gear train from the converter turbine shaft. Shaft B is the output shaft, which can be locked to the final drive pinion by means of the forward clutch.

First gear. In the normal drive arrangement (with D selected at the quadrant) the forward clutch is engaged, resulting in the housing E tending to turn in the opposite direction, causing it to lock up on the sprag clutch. Power is applied to shaft A, transmitting through pinions 2, 4, 5 and 6 to the output shaft B. The compound gears 4 and 5 revolve in the bearing at 3. Free-wheeling in this gear is possible when reaction at the sprag clutch is removed.

Second gear. Second gear is brought into action by applying the second gear brake band in addition to the forward clutch. This action locks the pinion 7 against rotation. Input from shaft A is now transmitted by the pinions 2, 4, 5 and 6, with the gear cluster orbiting round the stationary pinion.

Third gear. The second gear band is released and the third gear band applied, to hold the pinion 8 against rotation. Input from shaft A is now transmitted to shaft B via the pinions 2, 4, 5 and 6, with the gear cluster orbiting around the pinion 8.

Fourth (top) gear. This is brought into action by releasing the third gear band and applying the top-reverse clutch in addition to the forward clutch. This action locks the pinions 6 and 7 to the final drive pinion. No relative movement can now take place within the gearset, and the assembly revolves as one unit.

Reverse gear. This entails the application of the reverse band to hold the housing E against rotation. The top-reverse clutch is applied, while the forward clutch is released. Power is transmitted from shaft A to shaft B via the pinions 2, 4, 5 and 7, rotating the final drive gear in the opposite direction to the input shaft.

What ratios are provided with this gearset?

In the Mini, pinions 2, 6, 7 and 8 have 29 teeth, pinion 5, 13 teeth, and pinion 4, 35 teeth.

First and reverse have the same reduction ratio of 2.69:1.
Second gear 1.845:1
Third gear 1.46:1
Fourth gear 1:1

What is the significance of the numbers 1, 2, 3 and 4 on the selector quadrant of the Automotive Products transmission?

This is a manual select control for use as a normal gear-change. No up- or down-shifts are possible in any of these positions. Automatic changes occur only when D is selected.

How can the gear ratios of the Automotive Products transmission be calculated?

In the Mini transmission, the gear wheels 2, 6, 7 and 8 as shown in Fig. 71 have 29 teeth; the compound gear wheels 4 and 5 have 35 and 13 teeth respectively.

First gear. In first gear, with the casing E locked by means of the sprag clutch F, the gear ratio between shaft A and shaft B is the same as for a simple gear train.

$$\frac{N_a}{N_b} = \frac{35 \times 29}{29 \times 13} = 2.69{:}1$$

Second gear. Assuming the input speed to be 1000 rev/min, find the speed of the arm, then find the speed of the output shaft.

To find the speed of the arm, consider the gear train made up of gears 2, 4, 5 and 7.

$$e = -\frac{29}{35} \times \frac{13}{29} = -\frac{13}{35}$$

$$e = \frac{N_7 - N_a}{N_f - N_a}$$

$$-\frac{13}{35}(1000 - N_a) = 0 - N_a$$

$$-13\,000 + 13\,N_a = 0 - 35\,N_a$$

$$48\,N_a = 13\,000$$

$$\text{Speed of arm } N_a = \frac{13\,000}{48} = 270.8 \text{ rev/min}$$

Now consider the train as composed of the gears 2, 4, 5 and 6.

$$e = +\frac{29}{35} \times \frac{13}{29} = +\frac{13}{35}$$

$$e = \frac{N_6 - 270.8}{1000 - 270.8}$$

$$+\frac{13}{35} \times 729.2 = N_6 - 290.8$$

$$270.8 + 270.8 = N_6$$

$$N_6 = 541.6 \text{ rev/min}$$

$$\text{Gear ratio} = \frac{1000}{541.6} = 1.845{:}1$$

Third gear. Again assuming an input speed of 1000 rev/min and solving in two parts as before.

To find speed of arm consider the train of gears 2, 4 and 8.

$$e = -\frac{29}{35} \times \frac{35}{29} = -1 = \frac{N_8 - N_a}{N_f - N_a}$$

$$-1 = \frac{0 - N_a}{1000 - N_a}$$

$$-1\,(1000 - N_a) = 0 - N_a$$

$$-1000 + N_a = 0 - N_a$$

$$2N_a = \frac{1000}{2}$$

$$N_a = 500 \text{ rev/min or } 2{:}1 \text{ ratio}$$

To find the speed of the output shaft, consider the train of gears 2, 4, 5 and 6.

$$e = +\frac{29}{35} \times \frac{13}{29} = +\frac{13}{35}$$

$$+\frac{13}{35} = \frac{N_6 - 500}{1000 - 500}$$

$$+\left(\frac{13}{35} \times 500\right) = N_6 - 500$$

$$N_6 = 685.7$$

$$\text{Gear ratio} = \frac{1000}{685.7} = 1.46{:}1$$

Top gear. This is a direct drive of 1:1.

Reverse gear. For reverse motion the differential housing is held by means of a brake band. The gear train is the same as for first gear, numerically $\dfrac{35 \times 29}{29 \times 13} = 2.69{:}1.$

How are gear changes arranged in automatic boxes?

Changes are usually effected by engagement and/or release of the brake bands or clutches. In the gearsets discussed in previous questions the forward automatic drive D is brought into engagement by applying a multi-plate clutch. This is similar to that in Fig. 27, which remains in engagement in all forward gears. First gear engagement is caused by the reaction on the sprag clutch. For intermediate changes, brake bands are applied by hydraulic servos, while top gear in all gearsets involves the application of two clutches, with all the brake bands released.

How are the changes signalled?

Up- or down-shifts are signalled by spring-loaded valves moving to direct oil to the necessary clutches or brake bands. Movement of the valve is controlled by the fluid pressure in the hydraulic system, namely the line pressure, governor pressure and throttle pressure. A typical shift valve is shown in Fig. 72. The force exerted on the valve against the spring load is regulated by the different pressures from the above circuits applied to the required lands on the valve.

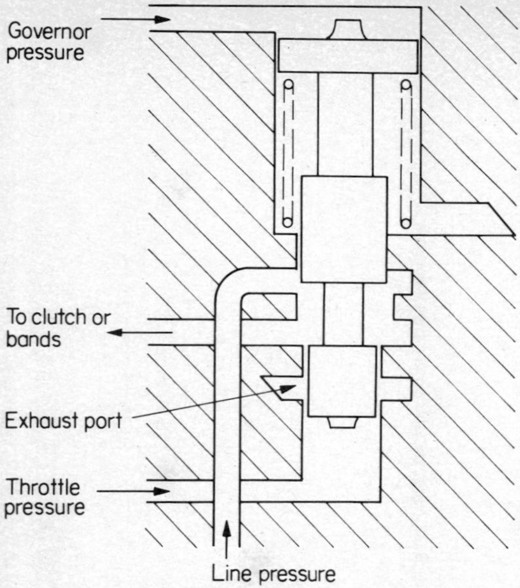

Fig. 72. Typical shift valve before up-shift

What is meant by line pressure?

This is pressure created by the oil pump in the transmission casing to supply the converter and a regulated pressure to shift and other valves in the circuit.

What is governor pressure?

To obtain a pressure based on speed, a centrifugal governor is introduced to regulate the pressure in this circuit.

What is throttle pressure?

A spring-loaded valve in the hydraulic circuit is connected to the engine throttle valve so that the movement of the throttle, in relation to power output, will regulate pressure in this line. A down-shift valve is also incorporated together with the throttle valve to provide for 'kick-down' changes.

What is a 'kick-down' change?

If the throttle pedal is 'kicked down' to the floorpan, the cable connecting the throttle with the gearbox causes the down-shift valve to move to the full extent of its travel to provide a change to the next lower gear, giving increased acceleration capacity as and when required.

Can a car fitted with an automatic transmission be tow-started?

This will depend on the number of oil pumps in the transmission's hydraulic system. Transmissions with only one oil pump, at the front, driven from the converter casing cannot be tow-started. Transmissions fitted with a rear pump or an auxiliary starting pump in addition to the front pump can be tow-started.

Can a car fitted with an automatic transmission be towed?

If long distance towing is necessary, and the internal lubrication system is not known, it is safest to disconnect the propeller shaft. Without an oil pump connected to the output shaft, lubrication within the transmission could break down and serious damage result.

Can faults in automatic transmissions be diagnosed?

This depends on the method of operation within the transmission. For example, if drive is not available when selected, it is possible that the sprag or one-way clutch is not functioning. If drive is available when first gear is locked by means of the brake band, this proves that the one-way clutch is not functioning. If the application of brakes and clutches is examined in each quadrant position, a logical sequence of testing by elimination will emerge.

Are there any other forms of automatic drive?

Throughout the history of the motor car, numerous attempts have been made to produce an acceptable variable speed drive. The most successful transmission of this type, as fitted to a motor car, is the Vee-belt drive as used by Van Doorne's Automobielfabriek Eindhoven, Daf Cars.

How is the speed ratio varied in this form of drive?

The belt pulleys have a fixed flange and a movable flange, so that the effective diameter of the pulley can be varied between limits. The position of the movable flanges on the drive pulleys is controlled by bob weights and engine vacuum. The driven pulley flanges are simply spring loaded (Fig. 73), and automatically adjust themselves to size according to the length of the belt. The fixed flanges of the driving and driven pulleys are on opposite sides to ensure alignment at all times.

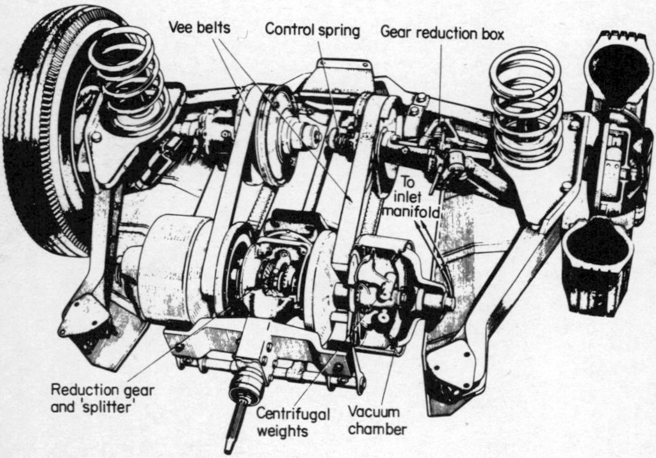

Fig. 73. Automatic transmission (Daf)

What arrangements provide for a neutral or out-of-gear position?

From the clutch (Fig. 28) the drive is taken to a bevel gearbox, with direct, neutral and reverse gear positions, selected by means of a small lever. The two side gears (27 teeth) are mounted on the cross-shaft, with a floating muff coupling between them. The muff coupling is splined to the shaft, so that movement from side to side will lock the alternative bevel gears to the shaft by means of the dogs on each end, for forward or reverse drive. The coupling is shown in Fig. 73 in the neutral position. The gearbox is sometimes referred to as the 'power-divider'.

Why is reduction gearing employed at the drive axle?

This is done to minimise the torque transmitted by the belts. The first stage reduction 16/27 is also limited on this account.

What factors govern the ratio selected by the variomatic transmission?

Engine speed acting through the bob weights, the vacuum applied from the inlet manifold, or the reservoir, to the chambers of the primary pulleys, and the 'load sensitivity' of the belt drive.

What is load sensitivity?

Load sensitivity is a feature of all Vee-belt drives having spring-loaded flanges. It arises from the fact that the tension in the loaded side of the belt opposes the rotation of the driving pulley, while causing rotation of the driven pulley. Any increase in tension ratio between the loaded and unloaded sides of the belt, owing to a rise in the applied torque or the resistance to motion, causes the driving belt to be pulled further into the driving pulley than the driven one, so the ratio is automatically raised; conversely a reduction of tension ratio or its reversal on overrun results in a lowering of the ratio.

How does the transmission operate?

There are five phases of operation, which can be followed by reference to Fig. 74.

When the throttle butterfly is in the idling position, the connection of the three-way vacuum valve to the reservoir is shut off, the other two connections being linked by passages in the valve. Under these conditions both chambers 1 and 2 of the primary pulleys are at atmospheric pressure. This condition applies through small throttle openings so that changes of ratio will be governed by the bob weights and the changes in tension ratio of the belts.

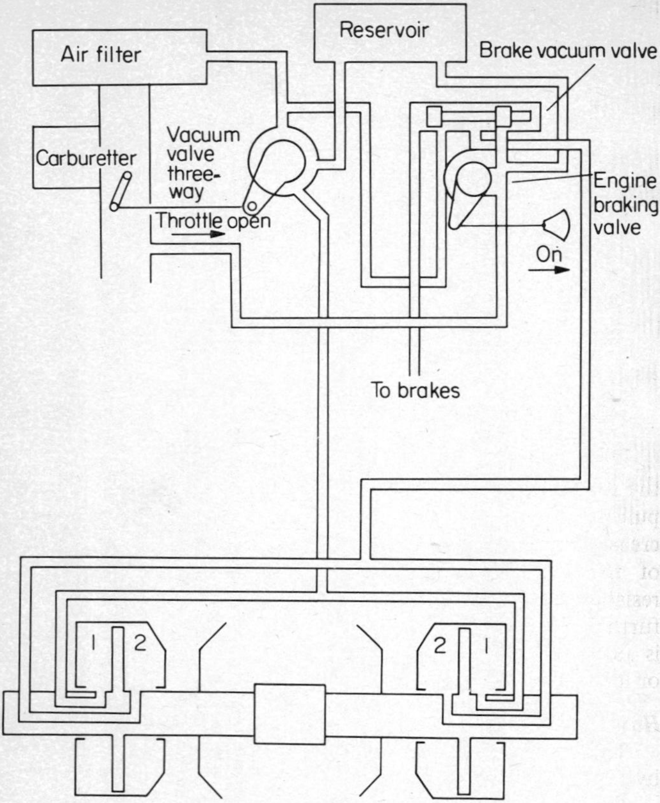

Fig. 74. Daf control system

Beyond a quarter throttle opening the connection to the air filter from the three-way vacuum valve is closed, the other two being now linked. Manifold depression is now applied to the outer chambers 1 of the pulleys by way of the non-return valve and the reservoir. The inner chambers 2 are at atmospheric pressure, and a pressure differential thus exists across the diaphragms. The differential assists the centrifugal action of the bob weights, causing the ratios to be reduced. This is known as the 'overdrive' condition, resulting in relatively low engine speeds for cruising.

Because of the non-return valve, the degree of vacuum in the outer chamber 1 remains constant, while that in the manifold is varied by throttle movement. It follows that, as before, changes in ratio are effected, to meet varying conditions of speed and load, by the bob weights and the tension of the belts.

When the throttle is opened beyond the three-quarter position, the reservoir is cut off and the outer chambers 1 of the pulleys are reconnected to the atmosphere, via the air filter. The pressure differential across the diaphragm no longer exists and there is a change to a higher range of ratios. This effect is analogous to the 'kick-down' of a normal, stepped automatic transmission.

When the accelerator pedal is released, the system reverts to the first stage of operation, and the transmission changes down gradually as the speed falls. However, because of the overrun condition of the belts, the ratio adopted at any given road speed is lower than it would be if the engine was pulling, so the braking effect is relatively small.

Increased braking on the overrun is available at any time by actuating the engine braking valve. The makers recommend that the valve should be used below 50 km/h to avoid over-speeding the engine. Operation of the brake control valve causes the inner chambers 2 of the pulleys to be connected to full engine vacuum, while the outer ones are subjected to atmospheric pressure through the three-way valve and the air filter. This causes a pressure differential across the pulley diaphragms to oppose the action of the bob weights, urging the pulleys towards the high ratio position.

Light braking up to about $0.2g$ has no effect on the control system, but when the hydraulic pressure exceeds a pre-determined value, it moves the piston of the brake vacuum valve against the resistance of the spring. This movement connects the inner chambers 2 of the pulleys with engine vacuum, so that the brakes are assisted by the high degree of engine braking.

Is there any differential action as the car negotiates a corner?

The load sensitivity of the belts provides a limited amount of differential effect. When the vehicle is put into a curved path, the outside rear wheel tends to accelerate, and consequently, for a given period, the torque to be transmitted by the belt is reduced. The resulting fall in the tension ratio causes the effective diameter of the driving pulley to increase, so the drive ratio on that side is lowered slightly. Since the opposite effect occurs at the decelerated wheel, no circumferential slip will occur during normal cornering.

What happens if one wheel loses traction on a slippery surface?

The reduced resistance to motion will cause the drive ratio on that side to be lowered. As a result, some slip will occur, but without affecting the drive to the other wheel: the reduced output torque that results from the higher speed of the wheel at once reduces the tendency to spin. This effect also applies if traction is lost by both wheels simultaneously.

10

DRIVE ARRANGEMENTS

What is the 'Hotchkiss drive'?

This is a popular drive system (Fig. 75) for front-engine, rear-wheel drive cars, using an open propeller shaft with two universal joints, driving a 'live' rear axle, which is attached to the chassis frame or body sill, with two semi-elliptic springs.

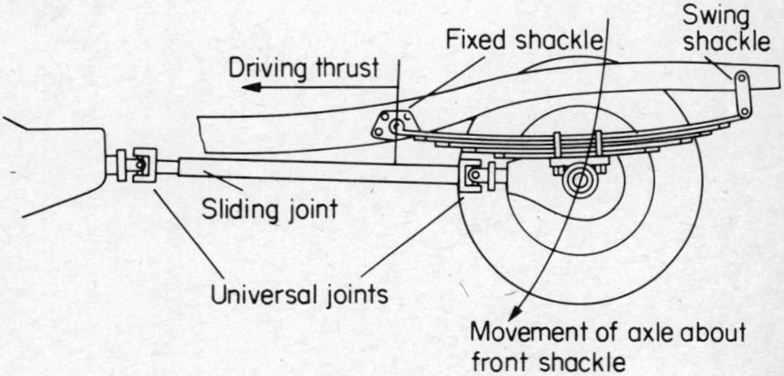

Fig. 75. Hotchkiss drive

What are the main features of the Hotchkiss drive?

The forward thrust of the road wheels is transmitted from the rear axle casing, through the front half of the road springs to the chassis frame or body sill; therefore the system must have semi-elliptic springs, rigidly attached to the rear axle casing and anchored to a fixed shackle at the forward end. The road springs must be made stiffer than other types to take this thrust and also resist the torque reaction set up by the drive.

As the car moves along the road the deflection of the road springs will cause the axle centre to move about the fixed shackle in the arc of a circle. It follows that the length of the drive from the axle centre line to the gearbox will be constantly varying. To accommodate this a sliding joint is incorporated in the propeller shaft construction, or within the gearbox extension.

Owing to the torque reaction (i.e. the tendency for the pinion to climb round the crown wheel) it is necessary to fit two universal joints to prevent any bending of the propeller shaft.

Is there a drive arrangement enclosing the propeller shaft?

Yes. This is sometimes called a 'torque tube drive', as the tube enclosing the propeller shaft resists the torque reaction of the pinion (see Fig. 76).

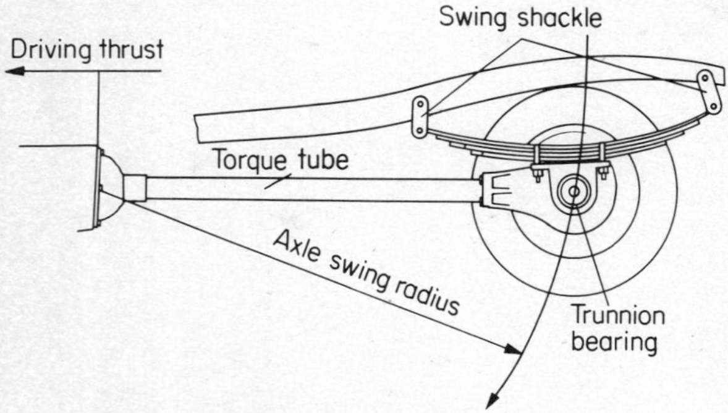

Fig. 76. The torque tube drive

What are the main features of the torque tube drive?

The tube is fixed rigidly to the rear axle casing, so that the axle centre moves in the arc of a circle, about the ball joint at the rear of the gearbox. This ball joint also houses the one universal joint used in the arrangement. The laminated road springs are mounted on swing shackles at both ends, which necessitates a trunnion block bearing at the point of attachment of the spring to the rear axle. A number of alternative attachments of the axle to body or chassis are possible, e.g. one laminated spring can be fitted transversely across the axle.

How is the forward driving thrust applied to the vehicle?

The driving thrust is applied at the rear of the gearbox, so that the gearbox or power plant has to be anchored to the chassis frame, or body, by means of straps. In this form of drive the vehicle can be said to be pulled along rather than pushed, as in the Hotchkiss drive.

Is there any other form of drive?

On long wheelbase vehicles the torque tube drive may be used at the rear axle, imparting the driving thrust to the chassis frame at a point, say, half-way along the vehicle. At this support point the drive changes to an open propeller shaft as in the Hotchkiss drive, to connect it with the power unit.

In front-engine, front-wheel drive, and in rear-engine, rear-wheel drive cars, the open drive shaft with two universal joints is often used, but swing axles (a variation of the torque tube drive) are in evidence, e.g. the Volkswagen (Beetle).

What are universal joints?

A universal joint connects two shafts, so that the second shaft is free to move away from the horizontal axis of the first shaft (or towards it if initially set at an angle away from it) during the transmission of power. Some types of joint, including the flexible disc, ring joints, cross joints and pot joints, cause a fluctuation in speed of the second shaft during each revolution.

What method is used to identify the joint, as in 'cross' joints, etc?

The member connecting the two shafts may be in the form of a cross, as in the Hardy Spicer or Hooke's universal joint. Some connecting members are in the form of a ring, as in the Rotoflex coupling. The 'pot' joint refers to any joint built into a cylindrical metal body.

What causes the fluctuation in speed between the two shafts connected by the joint?

Consider two positions of a cross joint as in Fig. 77. In Fig. 77 (a) the cross is in a vertical plane, at right-angles to the input shaft. In Fig. 77 (b) the cross is at an angle to the input shaft. This angle is equal to the angle made by the output shaft to the horizontal. If the outer point of one leg of the cross is plotted, as on the surface of a sphere, it will be found that the distance moved through 90° of rotation is greater than if the two shafts rotated on the same horizontal axis. To pass over this distance the leg of the cross has to accelerate, and then decelerate back to normal rotational speed at the 180° or

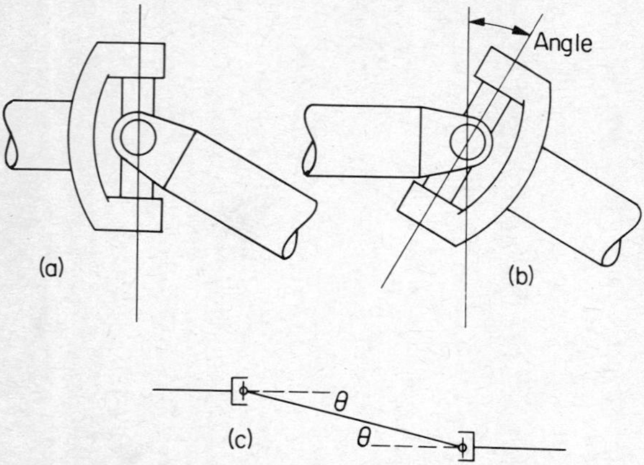

Fig. 77. Maximum and minimum positions of Hooke's joint

half-way mark. Thus the speed fluctuates every quarter revolution. The variation in speed is dependent on the angularity of the shaft. There is no variation at zero angle when both shafts are rotating on a mutual axis. As the angle increases so does the variation in speed.

How can the fluctuation in speed be corrected?

One way to prevent a variation in speed of drive is to use two universal joints, as in the Hotchkiss drive (Fig. 75), to form a three-shaft drive, with the forks of the propeller shaft in complete alignment one with the other. With this arrangement the acceleration of the first joint is cancelled by the deceleration of the second, so that the first (or gearbox) shaft and the third (or pinion) shaft run at a constant velocity, provided that each shaft joint is equally inclined to the other as in Fig.77 (c).

What happens to speed variation in the torque tube drive?

The speed variation exists in this form of drive, but if the power unit and the drive shaft are set, initially, at an angle to one another, it follows that when the car is laden the movement of the road springs will reduce the angle. The fluctuation in speed also reduces in unison as and when a mutual axis of rotation is approached.

Can this variation in speed be eliminated in a single joint?

The variation in speed is eliminated in the constant velocity joint. Any device in which the connecting (or centre) member maintains equal angles with the input and output shafts produces a constant velocity joint. To take the example of a three-shaft drive with two universal joints, it has been stated that constant velocity between the first and third shaft is possible provided that equal angles exist at each joint intersection. If the centre (or intermediate) shaft can in effect be reduced to zero length, as a development of a double Hooke's joint, a single constant velocity joint can be achieved.

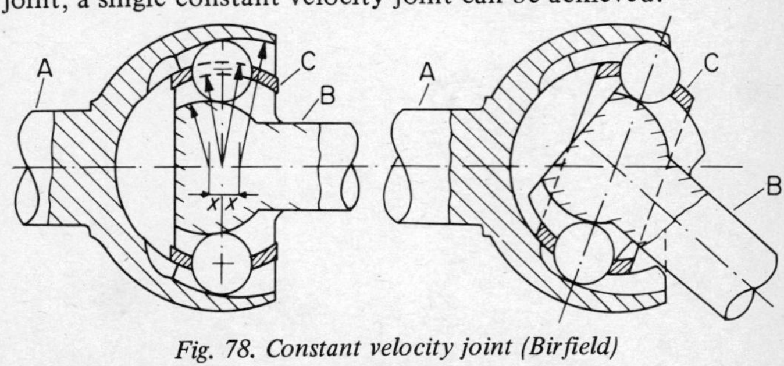

Fig. 78. Constant velocity joint (Birfield)

In the Birfield joint (Fig. 78) the intermediate shaft is replaced by the partially spherical shell C and the six ball bearings. Movement of the inner shaft B turns the ball bearings so that the cage assumes a position of half the angle between the shafts A and B. Thus the cage rotates in a plane making equal angles with the two shafts at all times to provide constant velocity between them.

Can worn universal joints be repaired?

In the earlier types of Hardy Spicer joints, repair kits, consisting of a centre spider (or cross) complete with needle roller bearings, bearing caps, oil seals and retaining circlips, were available. To dismantle the joint the existing circlips should be removed and the holes in the yokes accommodating the bearing caps carefully cleared of dirt, paint or underseal. To remove the caps, tap the yoke smartly with a copper hammer, adjacent to the bearing caps. This tapping will cause the bearing caps to lift out of the yokes. To reassemble, press in the bearing caps complete with needle bearings and oil seals, using brass or other soft metal buttons between the bearing caps and the vice jaws. In recent times some car manufacturers have sealed in the bearing caps, so that in the event of worn bearings a complete replacement shaft is the only answer.

In joints such as the rubber bushed Dunlop joint (Fig. 79)

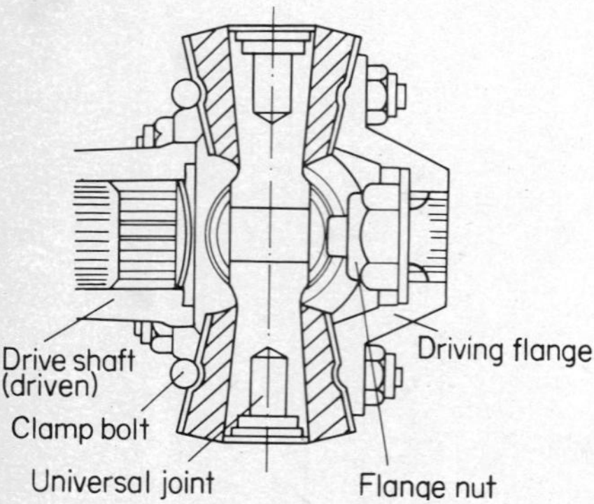

Fig. 79. Dunlop universal joint

and the Rotoflex coupling (Fig. 80) the centre unit is replaceable as and when required. Rubber joints require no lubrication and they also provide a cushioning and sound deadening effect.

Constant velocity joints of the Birfield type (Fig. 78) need a different approach. Suspected joints should be dismantled and inspected for wear. Worn joints should be replaced as a complete unit, as they use a special grease and should be sealed against the entry of dust and grit.

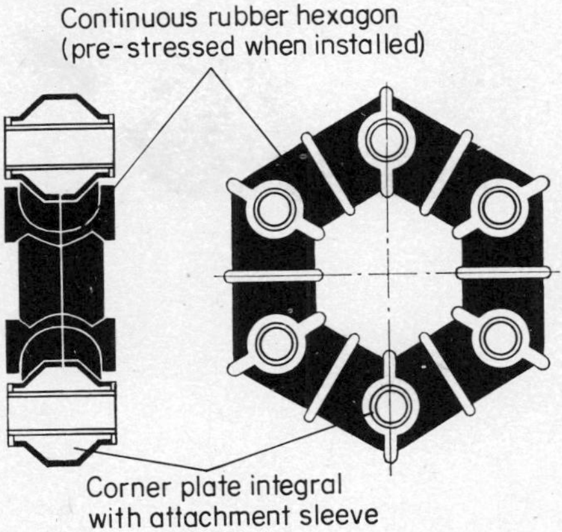

Fig. 80. Rotoflex universal joint (Dunlop)

What is the major role of the constant velocity joint?

The Birfield or similar type of joint is essential on front-wheel drive cars. The acceleration and deceleration of a normal joint would cause excessive wear on the tyres. The Birfield is a popular joint because of the wide angle through which it will operate.

What is the cause of vibration in a propeller shaft?

Vibrations can be set up by the shaft assembly being out of balance. If the shaft is built up on a sliding joint at the forward end, the mating parts will be marked for the correct balance position. When the sliding joint is inside the gearbox extension, the propeller shaft is a complete unit, usually carrying a balance weight in the form of a piece of steel spot-welded to the tube. Long propeller shafts are more susceptible to vibrations than short ones.

What is the reason for three universal joints in a given propeller shaft?

Three joints are often found in G.T. versions of some car models. The idea is to move away from the comparatively long shaft to avoid transmission vibrations. The positioning of the centre bearing and support is critical, so that the fluctuation of this joint is minimal.

11

REAR AXLES

What is meant by 'live' and 'dead' axles?

A 'live' axle consists of a hollow casing through which the power is transmitted, passing from the final drive to the differential housing and the differential gears, to the axle shafts and hence to the road wheels.

A 'dead' axle is a solid member attached to the suspension springs, with the road wheels mounted on suitable bearings at either end.

What is the 'final drive'?

The final drive is the reduction gear used in a live rear axle to turn the drive through 90°. The type of gearing used may be 'spiral bevel', 'hypoid', or 'worm and worm wheel'.

What are the average ratios for spiral bevel, hypoid, and worm and worm wheel reduction gears?

The spiral bevel is limited to about 5:1, the hypoid gear to about 7:1, while the worm and worm wheel can range from 10:1 up to 22:1, depending on the application.

What is spiral bevel gearing?

The final drive, i.e. pinion and crown wheel, have teeth which are curved throughout their length. If these curves were continued they would intersect at the cone centre of the gears (Fig. 81). The right-hand spiral is normally used, with the pinion mounted so that its centre line passes through the centre of the crown wheel. The spiral angle is the same for

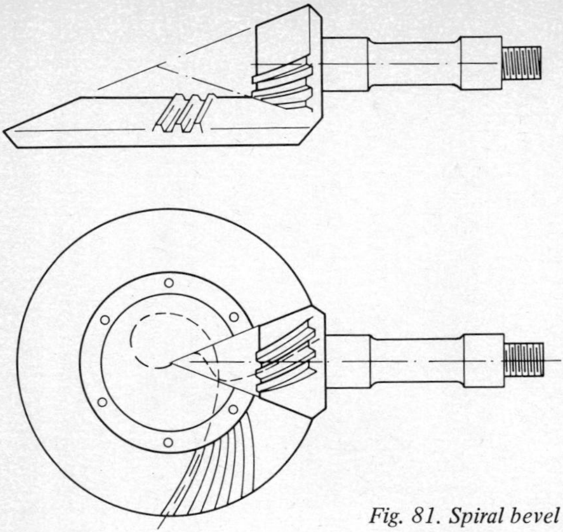

Fig. 81. Spiral bevel gearing

both pinion and wheel teeth, allowing movement of the pinion over a fairly wide range of positions for meshing purposes. More than one tooth is in engagement at any one time, contributing to smoothness in its action.

What is hypoid gearing?

The hypoid gear can be recognised by the offset given to the pinion shaft (Fig. 82). For clockwise (normal) engine rotation the offset is below the crown wheel centre line, which in turn allows for a lower transmission line from the gearbox in rear-wheel drive cars. In some transmissions the pinion shaft is above the crown wheel centre (see page 149). The pinion cone centre is determined by the amount of offset and the diameter of the gears. This results in the pinion itself being larger than those used in spiral gearing.

The spiral angle of the pinion teeth is greater than that of the crown wheel teeth, resulting in rolling and sliding between the contact surfaces during rotation of the gears. This form of gearing is smooth in operation, but requires special lubrication, owing to the high pressure that can exist from time to time between the teeth.

With hypoid gearing the pinion shaft can continue beyond the pinion so that it can be 'straddle' mounted between two bearings. This prevents deflection of the pinion shaft under extreme load conditions.

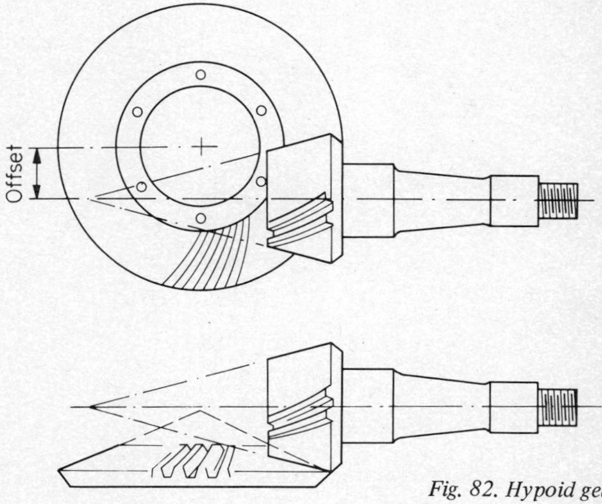

Fig. 82. Hypoid gearing

What is worm and worm wheel reduction gearing?

This form of gearing, shown in Fig. 83 as a double-start worm in part-section, permits the use of gear ratios in the order of 10:1 to 15:1 for use in heavy transport and passenger service vehicles. Some farm tractors have this type of gearing in their transmission lines in excess of 20:1. It is normally silent in operation, capable of withstanding heavy transmission and driving shock loads. It is also easy to assemble for correct tooth alignment.

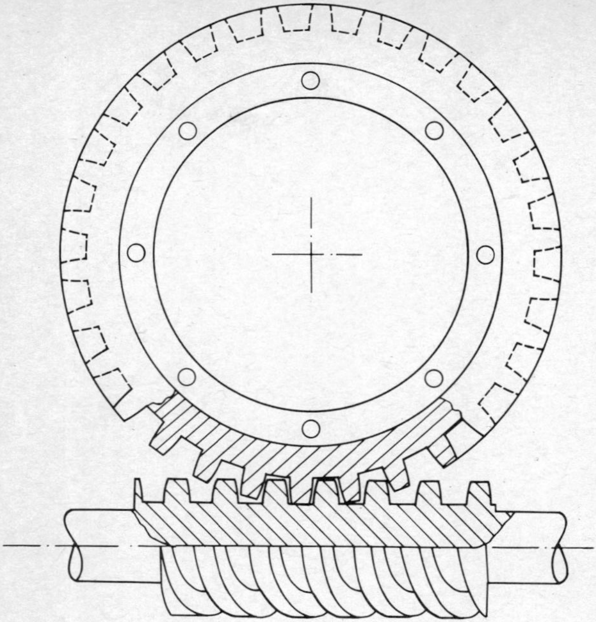

Fig. 83. Worm and worm wheel

A design advantage of this form of gearing is that it can be arranged with the worm above or below the worm wheel, depending on whether the vehicle is best suited for a high or a low transmission line. The position of the worm does not affect the transmission efficiency, and the ratios can be varied by having a number of different 'starts' on the worm.

What are 'starts' on the worm?

The word 'starts' refers to the number of helices on the worm. A single start is similar to the thread on a bolt. A double-start will have two helices at 180° phase angle, while a three-start worm will have three helices at 120° phase angle.

What is a differential gear?

This is the gear system which connects the two half-shafts at the centre of the axle casing (Fig. 84). The drive is transmitted from the crown wheel (or worm wheel) to the differential housing, hence to the drive pin and then divided between the differential gears to drive the axle shafts and road wheels.

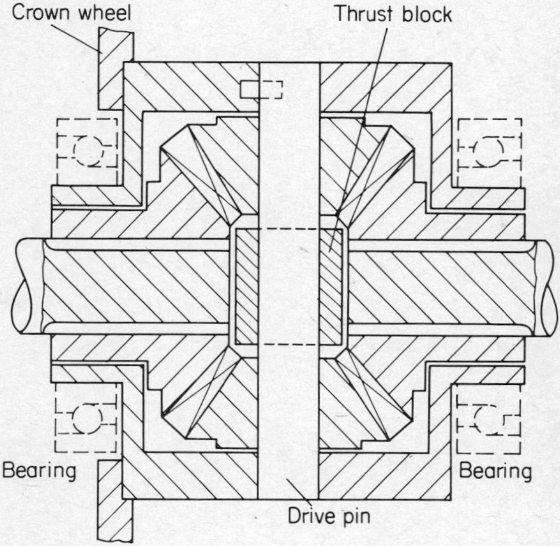

Fig. 84. The differential gear

What is the purpose of the differential gear?

It enables the road wheels to run at different speeds when negotiating a bend or corner. To take a right-angle turn the outer wheel must travel a greater distance than the inner wheel in exactly the same time; thus the outer wheel runs faster during these conditions.

How is the difference in speed accomplished?

First consider a vehicle running straight when the resistance to rotation on each road wheel is equal. The torque applied to the housing and the drive pin is equally applied to the axle shafts through each side gear, without the gears on the drive pin rotating about it. When the vehicle negotiates a corner, the resistance to rotation of the inner wheel is increased and the resistance of the outer wheel decreased. In this condition the gears on the drive pin run around the slow bevel to increase the speed of the outer wheel.

Does the torque on the axle shafts vary with the speed of rotation?

No, the torque applied to the drive pin is divided equally between the two shafts at all times.

What happens to the torque when one wheel loses its grip?

If a driving wheel slips on grease or ice, any torque applied to it will simply cause it to spin under these conditions.

The only torque that can be transmitted to the wheel will be that required to overcome the friction in the wheel bearing and its adhesion with the ground. As the differential equalises the torques transmitted, the same small torque is applied to the other wheel. This may have a perfect grip with the road but it cannot be utilised.

What is the gear ratio of a differential gear?

The ratio of any differential gear is 2:1, i.e. with one road wheel jacked up off the ground, two turns of the wheel will cause the crown wheel to make one revolution. In Fig. 85, the line on the left represents the distance travelled by the axle shaft at that side. The cross on the right indicates that this axle and side gear is at rest. For one complete revolution of the left-hand axle, the drive pin (represented by the circles) moves through half the distance, giving a 1:½ or 2:1 ratio.

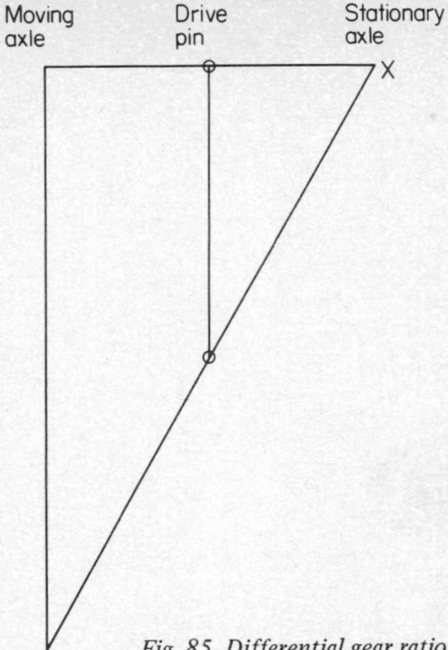

Fig. 85. Differential gear ratio

Can the loss of drive due to wheel slip be corrected?

One of the earlier methods of overcoming this difficulty was to fit a differential lock. With this arrangement, one or both of the differential wheels could be locked to the housing by means of dog clutches. The more recent method is the use of friction in what is generally called a limited slip differential.

What is a limited slip differential?

An example of this type of differential is shown in Fig. 86 The small clutches on either side of the differential gears are initially spring-loaded by one of the clutch plates being dished as in a Belleville washer.

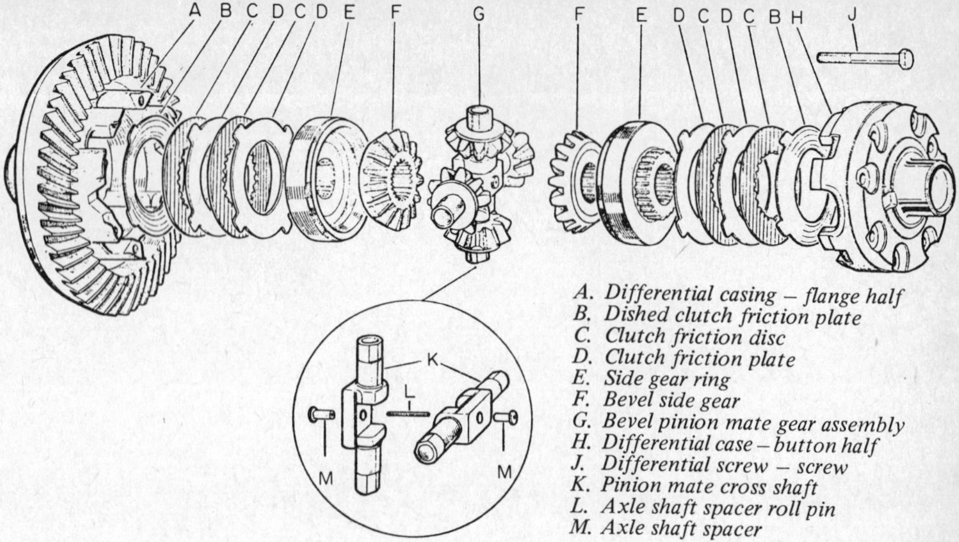

A. Differential casing – flange half
B. Dished clutch friction plate
C. Clutch friction disc
D. Clutch friction plate
E. Side gear ring
F. Bevel side gear
G. Bevel pinion mate gear assembly
H. Differential case – button half
J. Differential screw – screw
K. Pinion mate cross shaft
L. Axle shaft spacer roll pin
M. Axle shaft spacer

Fig. 86. Limited-slip differential (Thornton Power-Lok)

When the gear is driving there is a separating force on the wheels which is directly proportional to the torque being transmitted. This force is applied to the clutch pack in addition to the initial loading. On many occasions this friction set up within the housing is sufficient to prevent slip and loss of drive. In extreme conditions of increased loading the 'spider' or cross moves up the ramp of the Vee-slot in the differential case, increasing the load on the clutch to apply the greatest driving force to the wheel that has the best grip.

How are rear axles classified?

Live rear axles are usually classified as fully floating, three-quarter floating, or semi-floating arrangements. The term is governed by the type of bearing used, its location within the arrangement and the loads to which the axle shaft may be subjected.

What are the loads that the axle shaft could carry?

Consider any rear axle under static conditions, supported by the wheels and loaded at the two springs by the vehicle weight and its load. The arrangement is a simply supported beam carrying two concentrated loads and is therefore subjected to bending moment and shear force. On cornering the axle shaft may be called upon to resist bending due to side forces acting at the wheel, or carry a compressive or tensile load, depending on whether in relation to the turn the inner or outer axle shaft is studied. Summarising, the axle shaft may carry bending or shear due to vertical static loading, bending or push-pull loading due to cornering, and under driving conditions it will always be subject to torsional stress.

What is a fully floating axle?

In this arrangement (Fig. 87) the axle shaft is subjected to driving torque only, under all conditions. The wheel hub is carried on two bearings mounted on the outside of the axle casing, the wheel load line passing between them with preference towards the inner bearing. The axle casing supports the vehicle as a beam, resisting bending and shear due to vertical

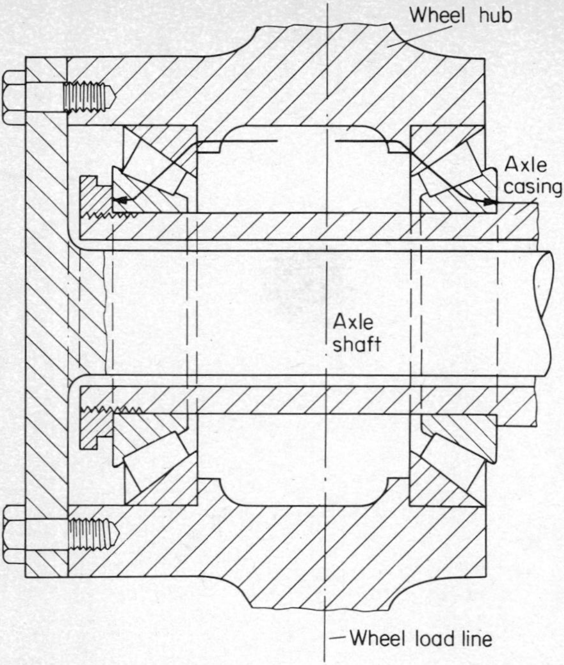

Fig. 87. Fully floating axle

loads. Bending due to cornering is resisted by the bearings and the axle casing, in the reverse order to that set up by the vertical loads. Push-pull is also resisted by the axle casing as indicated by the arrows.

In this arrangement the axle shaft can be removed without dismantling the wheel or jacking it off the ground.

What is a three-quarter floating axle?

The axle shaft carries driving torque only, when the vehicle is running straight, though when it negotiates a corner the shaft will be subjected to bending or axial loading depending on the type of bearing incorporated. Bending and shear due to vertical loading is eliminated by the wheel load line coinciding with the bearing centre line. In Fig. 88 a deep-groove ball bearing is employed. Stresses arising from vertical loads are eliminated by the wheel load line passing through the bearing centre line (there can be no bending if the support is directly under the load). Push-pull loads are resisted by the casing via the bearings as indicated by the arrows. The axle shaft will be subjected to bending in addition to torsion on cornering.

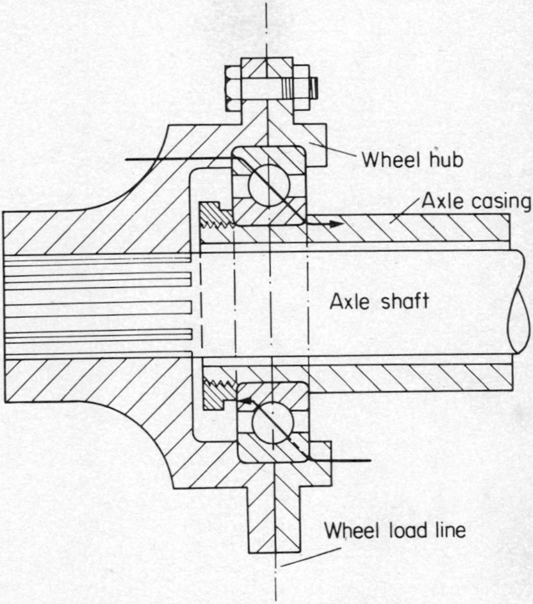

Fig. 88. Three-quarter floating axle

Can the loading of a three-quarter floating axle be altered by design?

Yes, a roller bearing mounted on the outside of the axle casing would resist bending due to cornering, while the push-pull loading could be transferred to the differential housing bearings. The wheel load line is designed to pass through the bearing, eliminating stresses due to vertical loading as before.

What is a semi-floating axle?

Any other design of rear axle wheel bearing arrangement is classified as semi-floating, even if it could in fact be assessed as a quarter-floating axle. In the arrangement shown in Fig. 89 the axle shaft is subjected to bending and shear due to the overhang that exists between the bearing centre line and the wheel load line. On cornering the axle will carry bending in addition to bend and shear due to vertical loads. The push-pull load is resisted by the axle casing as indicated by the arrows. It could be said that it is in fact a quarter-floating axle, as only one type of loading is applied to the axle casing.

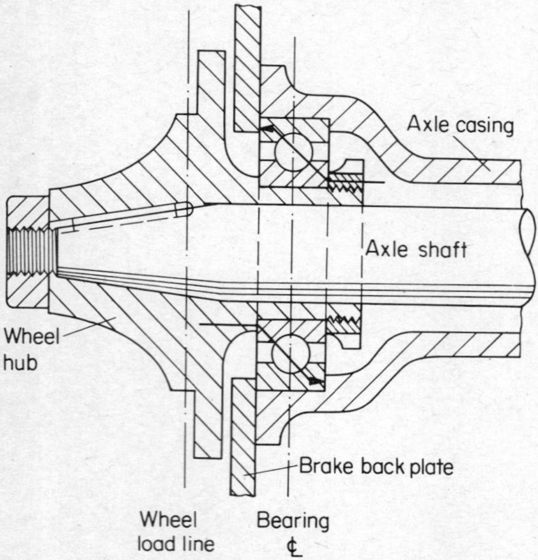

Fig. 89. Semi-floating axle

On what types of vehicles are these wheel bearing arrangements used?

Fully floating axles are generally found on light and heavy commercial vehicles. Three-quarter floating systems are used on relatively heavy cars, while semi-floating axles are used extensively on light cars.

How is the push or pull load resisted when it is transferred to the differential housing?

When this method is employed the differential side gear is locked to the axle shaft by means of a circlip, the alternative being a side gear forged to the shaft. With the aid of a thrust block, as in Fig. 84, the loading is transferred to the differential housing and hence to the bearings carrying the housing within the rear axle casing.

Can the differential be checked without dismantling the rear axle?

Yes, if both wheels are jacked up clear of the ground. With a low gear selected to hold the propeller shaft, rotation of one wheel will cause the opposite wheel to revolve in the other direction. The only wheels revolving in the centre of the axle casing are the differential side gears and the mating wheels on the drive pin. Any damage to the gear teeth should be felt if rotation of the road wheel is done by hand.

Can the vehicle be run in gear with one road wheel off the ground?

Provided that the axle is not fitted with a limited slip differential this test can be applied. It isolates each hub bearing and axle shaft, if the test is done on each side in turn. Any undue noise from the axle will be in the final drive and support bearing if the differential has been checked as described above.

Why is it not possible to test a limited slip differential in this manner?

This is the condition which the limited slip differential is designed to overcome, so that the differential side gear clutches would grip and drive the vehicle forward off the jack.

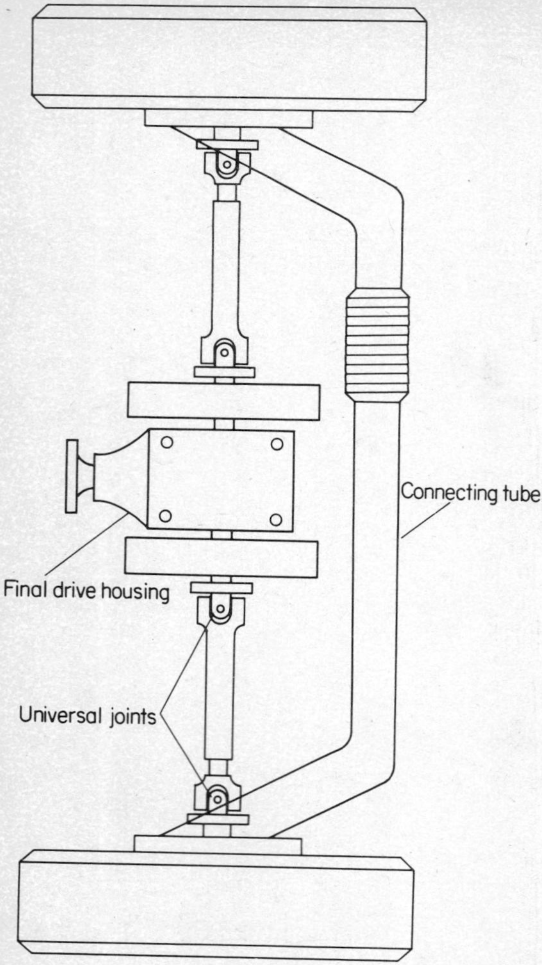

Fig. 90. De-Dion axle

How are the rear axle casings manufactured?

The design of the rear axle casing will depend on the load to be carried by the vehicle. For heavy transport and passenger service vehicles, one-piece drop-forged axles are used. For general car construction, the centre section of the axle starts as a piece of heavy section steel tube. The first operation is to slot the tube, and then a series of swaging tools form the circular housing to accommodate the final drive. This circular housing is now flanged to receive the attachment bolts of the final drive and cover plate. Two side tubes, with hub flanges attached, are then flash butt welded to the centre section.

An older method is to produce two halves of the casing by pressing and then welding together along the axis above and below the axle centre line to form a complete housing. The rear axle casing is often referred to as a 'banjo' casing owing to its resemblance to a double-ended banjo.

What other drive arrangements are in use?

The 'live' rear axle is a rigid member from wheel to wheel. A variation of this is the 'de-Dion' axle (Fig. 90) in which the torque tube and final drive housing are mounted direct to the car body or chassis. The drive to the wheel is accomplished by two short shafts incorporating two universal joints to obtain constant velocity. The two wheels are connected by the large diameter tube, so that the rise and fall of one wheel will alter the attitude of the opposite wheel. This system is therefore not independent.

What is an independent rear suspension?

Any system in which the rise and fall of one wheel does not affect the attitude of the opposite wheel can be said to be independent. The final drive housing is again attached to the body or chassis. The road wheels are mounted on two separate swing arms, which gives rise to the use of drive shafts incorporating two universal joints (Fig. 91). Note also the need for telescopic tubes, as indicated by the dust covers, on each drive shaft.

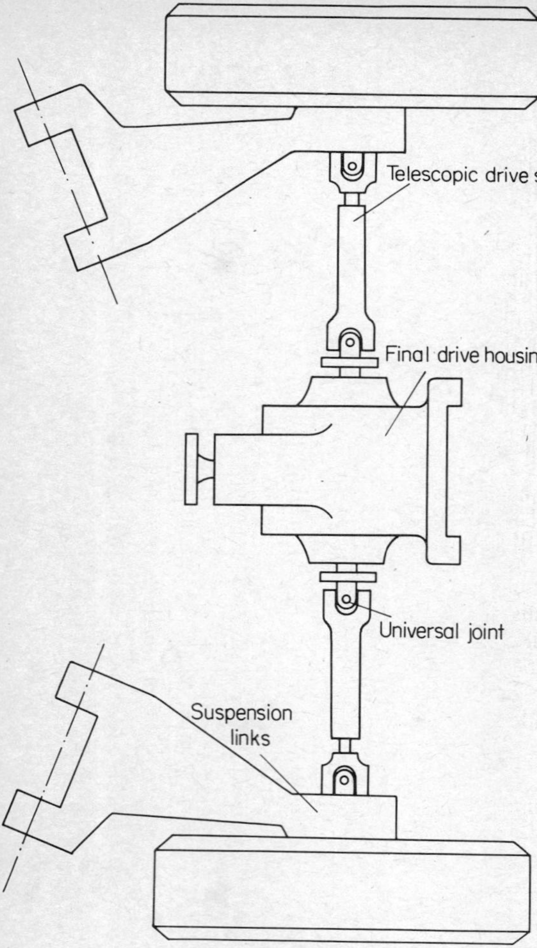

Fig. 91. Independent rear suspension

What is the difference between two-speed and double-reduction axles?

Two-speed axles have two separate ratios that can be selected at will by the driver. Double-reduction axles have two sets of reduction gearing, which are always in mesh. These are used on heavy transport and passenger service vehicles to carry the heavy loads and allow the design of low-loaders.

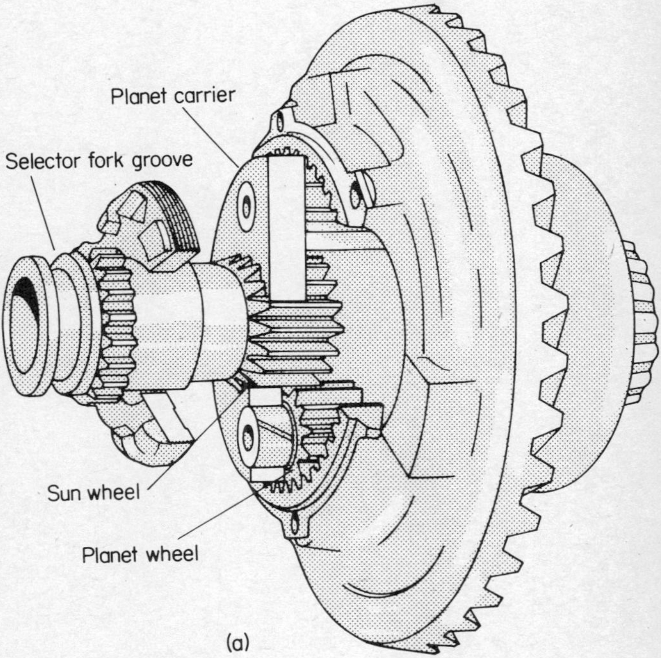

(a) Axle in high ratio

Fig. 92. Two-speed axle (Eaton)

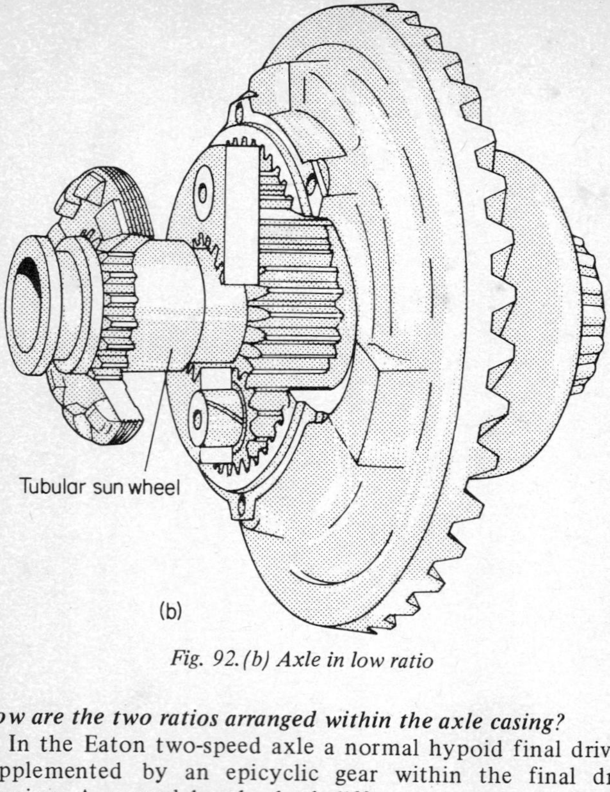

Fig. 92. (b) Axle in low ratio

How are the two ratios arranged within the axle casing?

In the Eaton two-speed axle a normal hypoid final drive is supplemented by an epicyclic gear within the final drive housing. A normal bevel wheel differential connects the two half-shafts. The sun wheel of the epicyclic train is tubular and sleeves over the near side axle shaft. The ring gear or annulus is machined within the crown wheel housing, with the planet wheels and carrier interposed between them.

How are the two speeds selected?

The tubular sun wheel is moved axially by means of a groove and selector fork to engage and disengage the two speeds. In the high ratio position (Fig. 92 (a)) the sun wheel is locking the planet wheels to their carrier, using the mating teeth as a dog clutch. The epicyclic gear is out of operation, and the differential functions as in a fixed ratio drive. When the tubular sun wheel is moved to the right (Fig. 92 (b)) the planet wheels are released, but the sun wheel is held by the outer dog clutch built into the axle casing. The drive is now in low gear, with the epicyclic gear in operation. The crown wheel and its housing, driving the planet wheels around the stationary sun wheel, turn the inner differential housing at a reduced speed.

How is the selector fork moved?

The selector fork can be moved by a mechanical linkage, but it is more usual to change speed by means of an electrical solenoid, or compressed air servo as shown in Fig. 93.

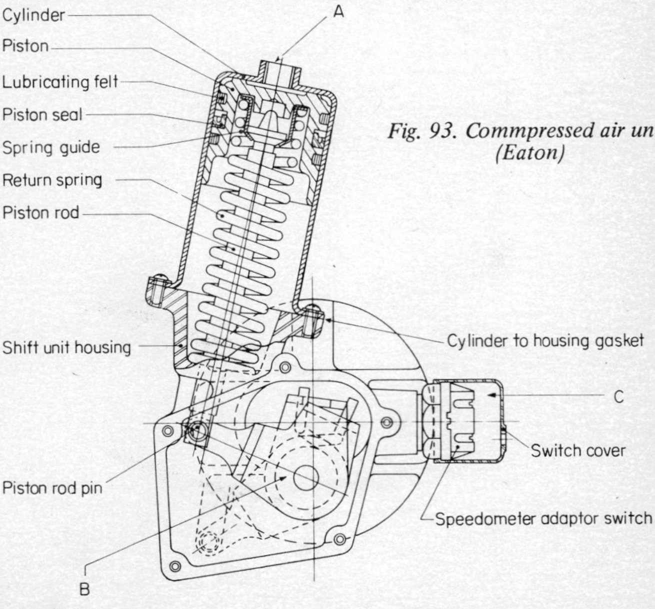

Fig. 93. Commpressed air unit (Eaton)

Is there a standard arrangement for double-reduction axles?

No. Once the direction of drive has been turned through 90° as in a hypoid bevel, the second reduction can be achieved by spur, bevel, or epicyclic gearing. In the construction of low-loaders the second reduction is often placed within the wheel hub.

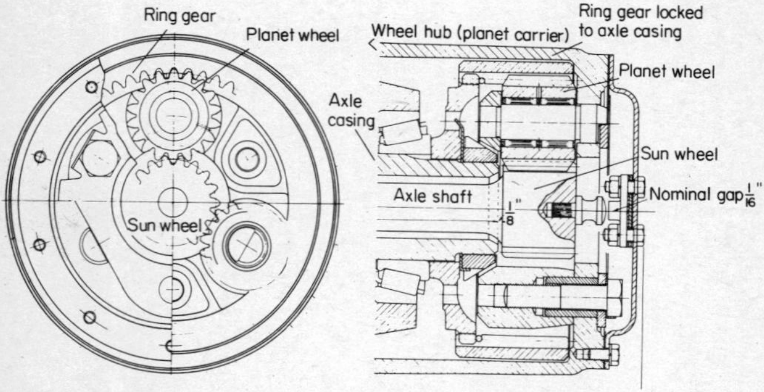

Fig. 94. Double-reduction axle (Albion)

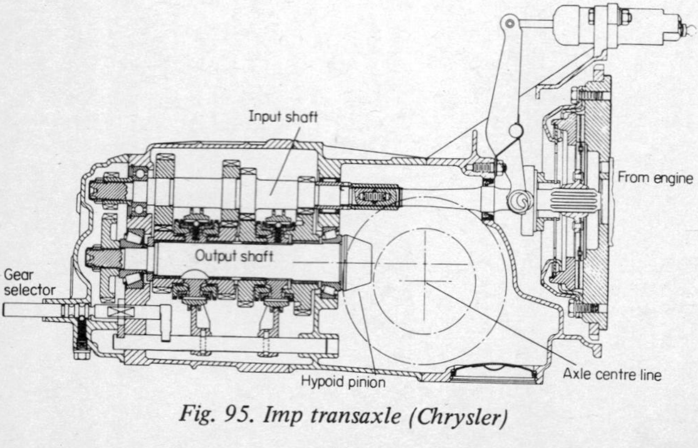

Fig. 95. Imp transaxle (Chrysler)

How is the epicyclic gear arranged within the hub?

The axle shaft usually carries the sun wheel, while the annulus or ring gear is machined within the axle casing flange (Fig. 94). The planet carrier is built into the wheel hub, with the planet wheels completing the drive. Power from the sun wheel drives the planets around the annulus to rotate the wheel hub in a reduction ratio.

What are transaxles?

In front-engine front-wheel drive and rear-engine rear-wheel drive cars, when the propeller shaft is eliminated the drive line from the engine to the road wheels becomes a transmission-axle or transaxle. The Hillman Imp transaxle (Fig. 95) is an example in which the hypoid gearing is used in the reverse order to a normal rear axle, i.e. the pinion is above the axle centre line. The gearbox is also opposite to normal, as the output shaft is the lower shaft carrying the synchromesh units. The overhung gears provide the means for reversing.

12

WHEELS AND TYRES

How are car wheels made?

Three types of car wheel are in general use, all with well-base rims. The most common is the steel disc (Fig. 96 (a)) using a steel rim circled and welded. This rim is then welded to the disc to form the wheel. There is no machining in its manufacture and the product is cheap, strong and of reasonable roundness. Some sports cars use the same basic rim attached to the hub by wire spokes (Fig. 96 (b)). This enables a certain amount of 'truing' to be done between the hub and the rim to prevent 'run-out'. The current trend is to fit cast wheels made in steel or light alloy, with the tyre seats turned to the correct form as specified by the type of tyre to be fitted (Fig. 96 (c)).

What is meant by a wheel size 5K-13?

This is a wheel size marking which is in three parts. The 5 represents the width in inches between the inside of the wheel flanges. K is the code letter for the rim section of K dimensions. The 13 is the rim diameter at the bead seat in inches. A list of coding letters together with the rim shapes is given in British Standards for the Automobile Industry. This information is also available in the latest edition of *Know about your tyres*, issued by the Automobile Association.

Who invented the well-base rim?

C. K. Welch devised the well-base rim in 1890, making it possible to build a tyre with inextensible wire in its edges, yet

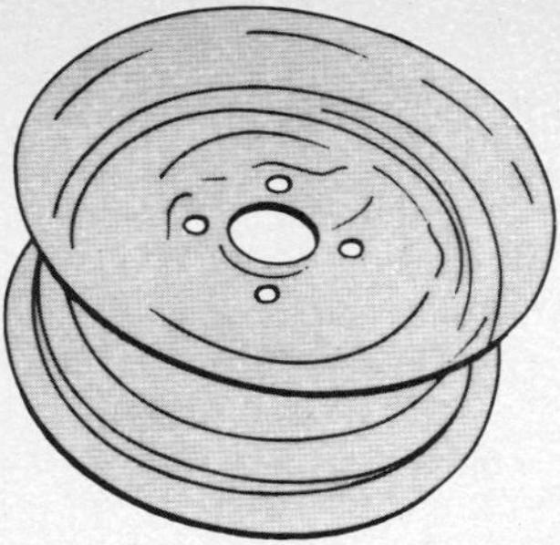

(c) Formula D2 wide rim steel wheel

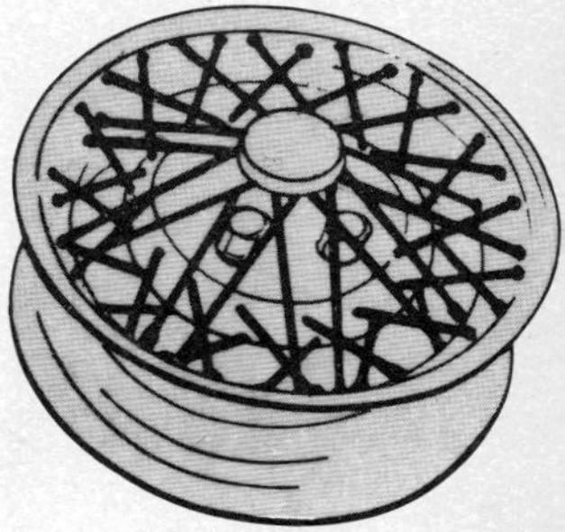

(a) Formula D3 bolt-on wire wheel

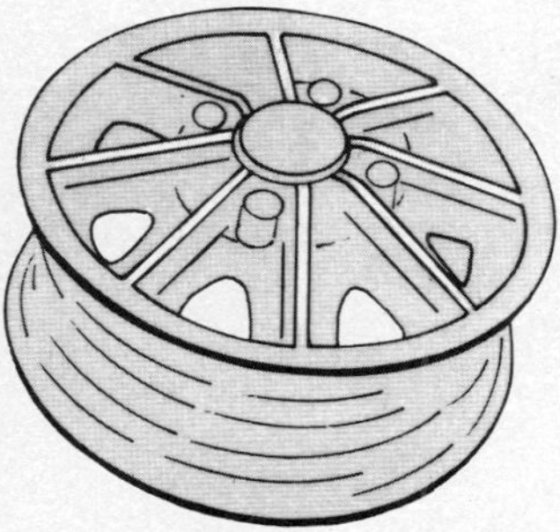

(b) Formula D1 light alloy wheel

Fig. 96. Car wheels (Dunlop Limited, from Information for Motor Vehicle Courses *booklets)*

simple to remove and refit by slipping the edge of the tyre into the well at one side, so that the tyre could be levered off the rim at a point diametrically opposite. This type of rim is still in use for cycles, cars and many other vehicles throughout the world.

Are the rims still made to the design of C. K. Welch?

Not quite. With the introduction of tubeless tyres and the need for an airtight seal at the tyre beads, the bead seats are made with a 5° taper so that when the tyre is inflated the bead is drawn up the taper by the air pressure to effect a good seal.

Has there been any further improvement in rim design?

In recent years, alterations have been made to make it more difficult for the tyre to be forced away from its rim during fast cornering. Fig. 97 shows a range of safety rims for cars. The long flat ledge prevents the tyre being forced into the well by locally applied force arising from cornering. The bead can only be forced into the well if the entire bead is pressed at the same time along the flat ledge and into the well. Some rim designs incorporate 'humps' to prevent tyre movement away from the rim.

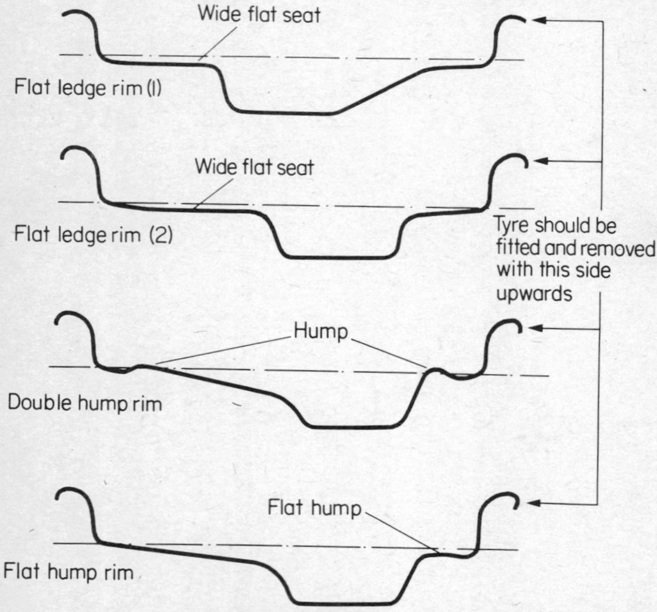

Fig. 97. Safety rims (Dunlop Limited, from Information for Motor Vehicle Courses *booklets)*

How is the driving force at the wheel determined?

The torque available at the rear axle shaft, divided by the effective radius of the wheel and tyre, determines the force available at the road surface.

What sets a limit to the force that can be applied?

The weight carried by the wheel and the coefficient of friction (adhesion) between the tyre and the road.

The formula $\dfrac{P}{W} = \mu$

where

P = Effort or driving force

W = Weight or load on the wheel

μ = Coefficient of friction or adhesion

shows that $P = \mu W$. Torque is equal to force × radius, hence the rear wheel torque = μWR, where R = Effective radius of wheel and tyre.

What steps have been taken to improve the tyre grip?

In the early 1960s a new type of tread rubber called 'high mu' came into use on cars. Fig. 98 shows the variation in coefficient of friction from dry roads to wet roads, with reference to high mu, synthetic and natural rubber treads. The high mu material, in the wet, has a grip nearly as high as that of natural rubber compound in the dry.

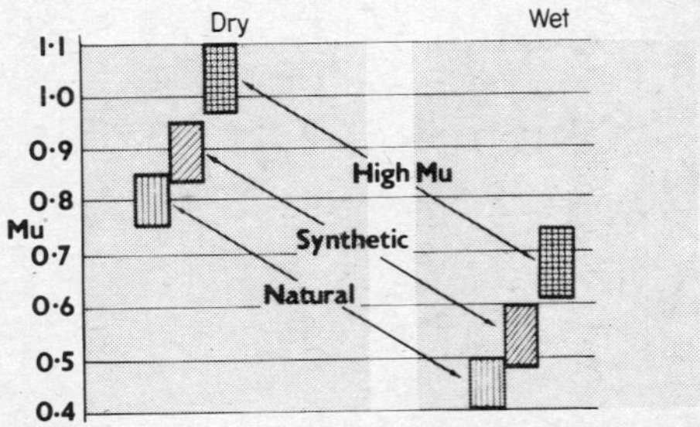

Fig. 98. Friction coefficients, dry-to-wet roads (Dunlop Limited, from Information for Motor Vehicle Courses *booklets)*

What is aquaplaning?

This is a condition of loss of grip owing to the depth of water on the road, and the inability of the tread pattern to feed the water away from the tyre contact area with the road.

Adverse conditions of lots of water and little tread pattern cause the wheel to plane along the surface of the water without turning.

How is the wheel lifted from the road surface?

A tyre under wet road conditions is shown in Fig. 99. Three zones exist in the contact patch of the tyre with the road. The first, 1, is filled with undisturbed water, while the second, 2, is partly cleared by the water being squeezed into the tread pattern, to be expelled from the sides of the tread and to the rear of the tyre. The third area, 3, shows the effect of the slots in the tread pattern mopping up the remaining water to give good contact with dry road.

As the speed rises the tyre may be unable to clear the water in time for the three stages to be completed and the whole contact area is held away from the road surface by a continuous film of water as the wedge finds its way under the wheel.

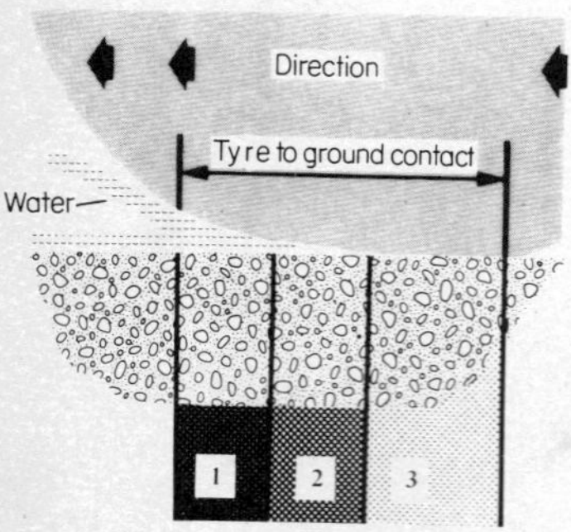

1 Area of unbroken water film, 'bulk' displacement area

2 Area of partial breakdown of water film, 'thin film' area. Road surface asperities penetrating.

3 Area of dry tyre to road contact, 'dry' area

Fig. 99. Aquaplaning (Dunlop Limited, from Information for Motor Vehicle Courses *booklets)*

How can the effective radius of the wheel be determined?

The effective radius will depend on the tyre pressure, so that measurement from the wheel centre to the road, with the tyre at the correct inflation pressure, is the most practical method. For general usage it can be found from the tyre size markings.

What information is given by tyre size markings?

Tyre sizes are based on the width of the tyre section and the diameter of the rim seat on the wheel it fits. This applies to all car and commercial vehicle sizes.

Examples.

Car size 5.20–10 A tyre 5.20 in wide on a rim of 10 in seat diameter
Overall diameter (5.20 × 2) + 10 + 20.4 in
Effective radius = 5.20 + 5 = 10.2 in

Radial tyres have section widths in millimetres. 165–15 is a tyre 165 mm wide on a rim of 15 in seat diameter. The rim diameter can also be in millimetres.

135–355 A tyre 135 mm wide on a rim of 355 mm seat diameter
Overall diameter (135 × 2) + 355 = 625 mm

$$\text{Effective radius } 135 + \frac{355}{2} = 312.5 \text{ mm}$$

Tyre widths in millimetres are an indication that the tyres are of radial construction. This can also be indicated by the letter R between the numbers, e.g. 165–R–13.

Truck size 10.00–20 A tyre 10 in wide on a rim of 20 in seat diameter
Overall diameter (10 × 2) + 20 = 40 in
Effective radius 10 + 10 = 20 in (508 mm).

What other information is included in tyre markings?

Additional letters given in the size code are speed markings: S for 'speed', H for 'high speed' and in radials V for 'very high speed', e.g. 5.50–S–12 and 185–HR–13. Table 1 shows the speed ratings for radial ply tyres.

Table 1. Speed markings for radial ply tyres

SR*	HR	VR
Up to 182 km/h	Up to 209 km/h	Over 209 km/h

*Also rating for reinforced radials up to 169 km/h.

Limiting speeds for cross-ply tyres vary according to rim diameter, as shown in Table 2.

Table 2. Variation in limiting speeds of cross-ply tyres according to rim diameter

Rim dia., in	*Normal,* km/h	S, km/h	H, km/h	V, km/h
10	up to 120	up to 153	up to 177	over 177
12	up to 137	up to 160	up to 185	over 185
13 and over	up to 153	up to 177	up to 200	over 200

What are 'low profile' tyres?

This is a trend in tyre design in which the height of the section is less than the width. A 70 series in which the height of the section is 70% of its width is already in production. Tyre markings for this series involve code letters, so that a typical marking reads D70–14. All tyres in this range are of radial construction.

How can radial ply tyres manufactured before this coding was introduced be identified?

Tyres without indication as to casing type were originally described as 'X'. This was followed by the 'Cinturato', the 'Max', 'SP', 'Unisteel' and others. Reference to manufactures' data may be necessary to prevent mixing of radial- with cross-ply tyres.

What is the difference between 'cross-ply' and 'radial-ply' tyres?

The basic difference is in the construction of the plies within the casing. The cross-ply tyre (Fig. 100) is constructed from fabric that has been cut so that the cords run at an angle of about 38° to the line of the bead. With this arrangement alternate layers of cord cross each other to form an included angle of about 45° to give rigidity to the casing. In the radial

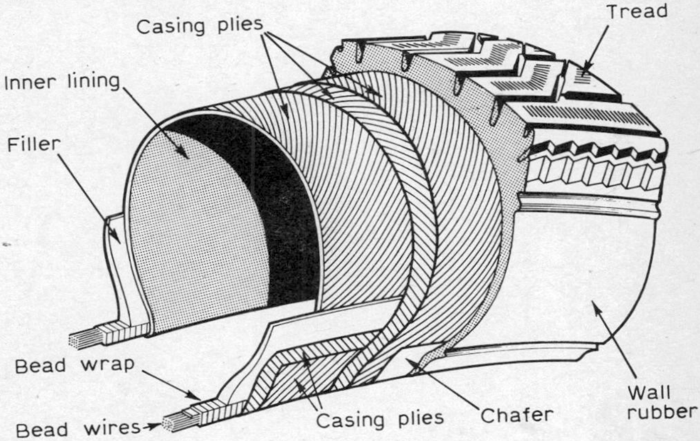

Fig. 100. Cross-ply tyre

ply tyre (Fig. 101) the casing plies are positioned with the cords running straight across from bead to bead on the line of a radius from the wheel centre. This form of construction gives a very supple casing to absorb vehicle movement. To prevent tread distortion, layers of cord are positioned above the radial plys to brace the tread. This bracing enables the tread to remain flat on the ground to give maximum grip under all conditions. Some tread bracings are made of steel.

Is there any other form of construction in general use?

Not in this country, or on the Continent. In the USA, tyre manufacturers produce a 'bias-belted' tyre. This is a cross-ply casing with tread reinforcement as in a radial tyre.

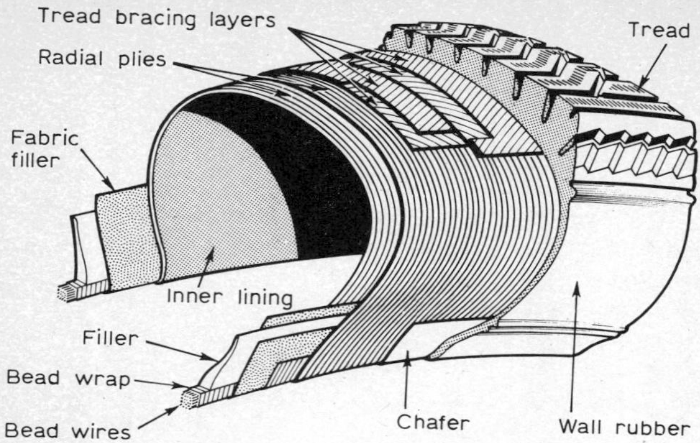

Fig. 101. Radial-ply tyre

For what reasons are radial-ply tyres fitted as original equipment on cars today?

The chief reason for the use of radial-ply tyres is that they are economical. They cost about 20% more to produce than a cross-ply tyre, but give 80 – 100% more mileage, depending on the type of radial used and how the car is driven. Cornering at higher speed is possible owing to the tread remaining in full contact with the road. It also follows that it provides an improved grip in wet weather. It is also more comfortable at speed, as its shock absorbing deflection is 25% greater than that of a cross-ply tyre. Changing from cross-ply to radial can show an improvement in fuel consumption of up to 5%.

What precautions must be taken when changing from cross-ply tyres, as original equipment, to radials?

It is dangerous to mix radial- and cross-ply tyres on a vehicle. The safest way is to fit radials all around, including the spare. It is important that the rim section is of the correct pattern to form an effective seal if the casing is of tubeless construction.

What are the legal requirements on the mixing of radial- and cross-ply tyres on a vehicle?

Types must not be mixed on the same axle, front or rear.

Radials must not be fitted to the front wheels only (even if the car has front-wheel drive).

Radials on the rear axle are permitted, except on fast sports cars (radials all around are better).

What precautions should be taken when using steel and textile radials on a vehicle?

Do not mix steel and textile radials on the same axle. If mixed tyres must be used, put two steel radials on the rear axle.

What are the legal requirements on tyre safety?

Tyres must be suitable for the use to which the vehicle is being put, and for use with the other tyres on it (see above).

Tyres must be properly inflated.

Tyres must be free from breaks or cuts more than 25 mm long and deep enough to reach the body cords.

Tyres must not have any lump or bulge caused by separation or partial failure of the structure.

Tyres must not have any portion of the ply or cord structure exposed.

Tyres must have a pattern of at least 1 mm in depth over three-quarters of the tread and extending around the entire circumference.

Is wheel balancing of practical importance for driving wheels?

On front-wheel drive cars it is of prime importance. On rear-wheel drive cars, static imbalance will cause 'wheel hop' or 'patter'. All wheels in balance, including the spare (for use in an emergency), is the ideal condition.

Why is wheel balancing necessary?

With the complicated structure of a modern tyre, it is impossible to obtain an even distribution of weight throughout, so that points of excess weight, under the action of centrifugal force, will cause out-of-balance effects.

What are out-of-balance effects?

The effects are caused by two forms of imbalance, static and dynamic. Static imbalance will cause 'wheel hop', 'patter' or 'tramp'. These terms explain the effect of centrifugal forces set up by the imbalance, tending to lift the wheel when the out-of-balance is at the top of the wheel, and to force it downwards when the out-of-balance is at the bottom of the wheel.

Dynamic imbalance occurs when two excess weights exist on either side of the tyre. The effect is to produce 'shimmy' in steered wheels.

How is the out-of-balance corrected?

Static and dynamic out-of-balance can be corrected by the addition of weights, at definite positions on the wheel rim. The placing of the weights and their magnitude is determined by the use of a balancing machine. Some machines balance the wheel off the car, others balance the wheel on the car. The relative methods will not be discussed, as they provide the answer to the problem in either case.

INDEX